8/12 002 00

EXPLORING URANUS

By Cody Keiser

Published in 2018 by
KidHaven Publishing, an Imprint of Greenhaven Publishing, LLC
353 3rd Avenue
Suite 255
New York, NY 10010

Designer: Deanna Paternostro
Editor: Vanessa Oswald

Photo credits: Cover, back cover, pp. 7, 19 (main) Vadim Sadovski/Shutterstock.com; pp. 4–5 ANDRZEJ WOJCICKI/SCIENCE PHOTO LIBRARY/Getty Images; p. 9 (main) shooarts/Shutterstock.com; p. 9 (inset) Tfr000/Wikimedia Commons; p. 11 Catmando/Shutterstock.com; p. 13 SkyPics Studio/Shutterstock.com; pp. 15, 19 (inset) Time Life Pictures/Contributor/The LIFE Picture Collection/Getty Images; p. 17 Jcpag2012/Wikimedia Commons; p. 21 Orange-kun/Wikimedia Commons.

Cataloging-in-Publication Data

Names: Keiser, Cody.
Title: Exploring Uranus / Cody Keiser.
Description: New York : KidHaven Publishing, 2018. | Series: Journey through our solar system | Includes index.
Identifiers: ISBN 9781534522893 (pbk.) | 9781534522817 (library bound) | ISBN 9781534522534 (6 pack) | ISBN 9781534522640 (ebook)
Subjects: LCSH: Uranus (Planet)–Juvenile literature.
Classification: LCC QB681.K395 2018 | DDC 523.47–dc23
Printed in the United States of America

CPSIA compliance information: Batch #BS17KL: For further information contact Greenhaven Publishing LLC, New York, New York at 1-844-317-7404.

CONTENTS

ICE PLANET

Uranus is the third–largest planet and the seventh planet from the sun in the **solar system**.

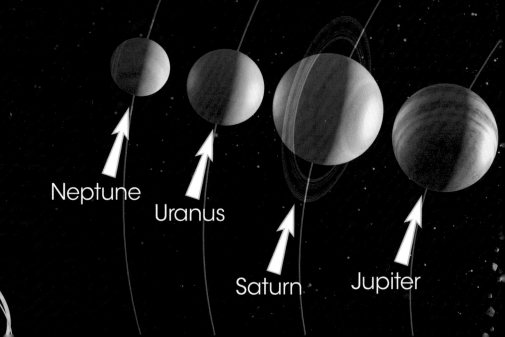

Neptune

Uranus

Saturn

Jupiter

It's about four times wider than Earth. Being so far away from the sun makes it very cold and icy.

Uranus most likely formed close to the sun and moved away from it about 4 billion years ago.

Mars

Earth

Venus

Mercury

sun

5

MOVING AND SPINNING

Like the other planets, Uranus **orbits** the sun. It takes Uranus 84 Earth years to orbit the sun once. It orbits slowly, but it spins much faster. It only takes 17 hours to spin around once!

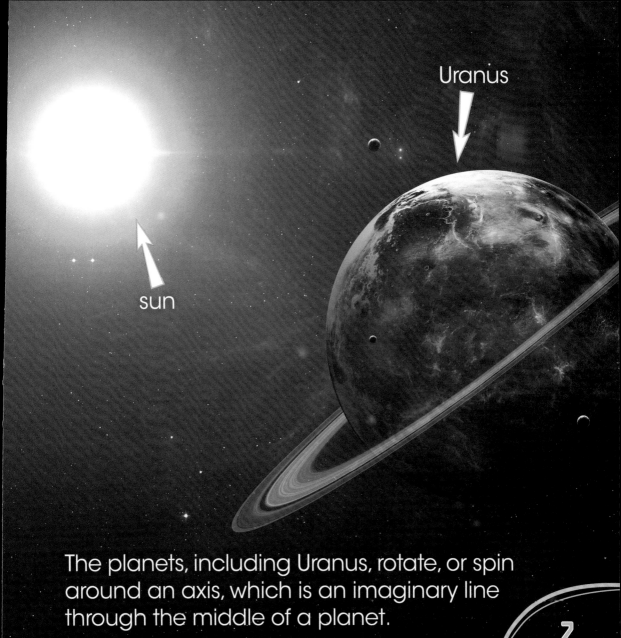

Uranus

sun

The planets, including Uranus, rotate, or spin around an axis, which is an imaginary line through the middle of a planet.

Uranus doesn't spin like most of the other planets. It spins in the opposite direction, which is called a retrograde rotation. Uranus is also different because it spins on its side!

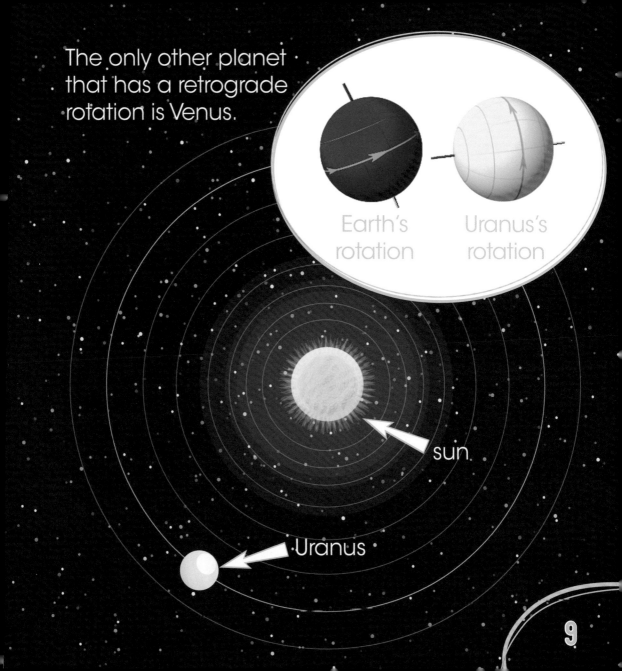

The only other planet that has a retrograde rotation is Venus.

Earth's rotation

Uranus's rotation

sun

Uranus

9

IT'S COLD!

Uranus has the coldest weather in the solar system. Icy clouds made up of the gases hydrogen, helium, and methane **surround** the planet. Methane makes Uranus look blue-green.

Frozen methane gives Uranus a cool blue color!

Scientists think a giant ocean made up of water, **ammonia,** and methane ices can be found beneath the clouds of Uranus. They also think there's a rocky core at the center of the planet.

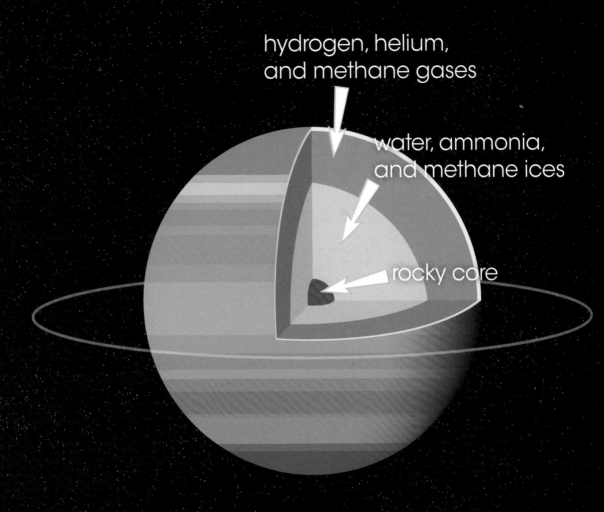

hydrogen, helium,
and methane gases

water, ammonia,
and methane ices

rocky core

The core of Uranus is
9,000 degrees Fahrenheit
(4,982 degrees Celsius).

RINGS AND MOONS

Thirteen rings surround Uranus. It has a set of inner rings and outer rings. The inner group has nine rings, and the outer group has four rings.

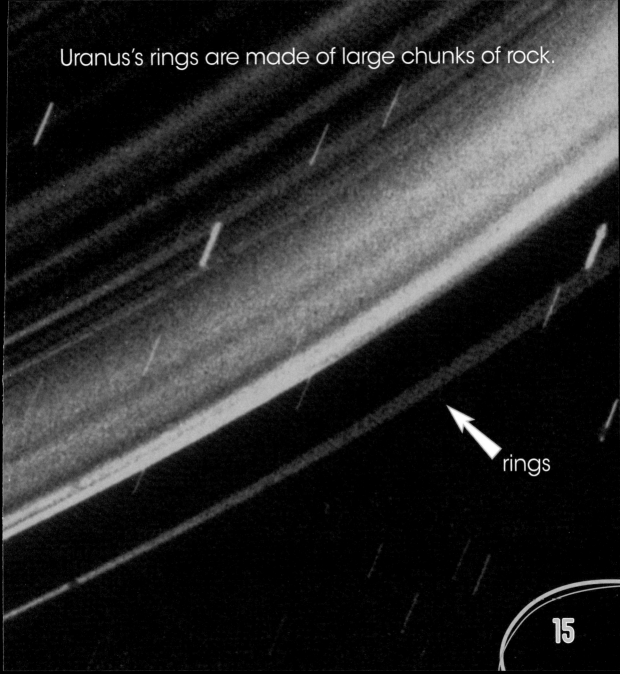

Uranus's rings are made of large chunks of rock.

rings

15

Uranus has 27 small moons. All the planet's inner moons are made of water, ice, and rock. Some scientists think the outer moons are **captured asteroids**.

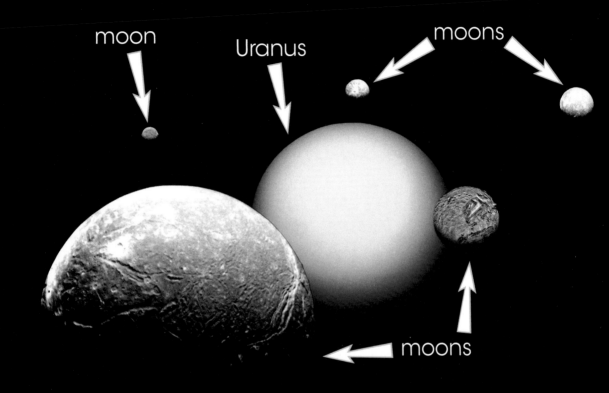

moon

Uranus

moons

moons

Uranus's moons were
first discovered in 1787.

STUDYING URANUS

The *Voyager 2* **probe** is the only spaceship that's visited Uranus so far. It launched in 1977 and traveled more than 1.8 billion miles (3 billion km). It took nine years to reach Uranus. The probe discovered many of the planet's rings and moons in six hours.

A spaceship couldn't land on Uranus because there isn't solid ground!

Uranus

Voyager 2 probe

Future **missions** to visit Uranus are planned for the late 2020s. Scientists can't wait to find out more about this cool planet!

If Uranus was the size
of a softball, Earth would
be about as big as a nickel.

Uranus

Earth

GLOSSARY

ammonia: A colorless gas with a strong smell.

captured asteroid: A large space rock caught in a planet's orbit.

mission: A definite task involving space.

orbit: To travel in a circle or oval around something.

probe: A vehicle that sends information about an object in space back to Earth.

solar system: The sun and all the space objects that orbit it, including planets and their moons.

surround: To be on every side of something.

FOR MORE INFORMATION

Websites

NASA Space Place: All About Uranus
spaceplace.nasa.gov/all-about-uranus/en/
NASA's website gives visitors a closer look
at Uranus.

National Geographic Kids: Mission to Uranus
kids.nationalgeographic.com/explore/space/
mission-to-uranus/#uranus-planet.jpg
This website presents fun facts about Uranus.

Books

Adamson, Thomas K. *The Secrets of Uranus*. North
Mankato, MN: Capstone Press, 2016.

Bloom, J.P. *Uranus*. Minneapolis, MN: Abdo Kids, 2015.

Glaser, Chaya. *Uranus: Cold and Blue*. New York, NY:
Bearport Publishing, 2015.

INDEX

Praise for

LIGHT & SHADE

"Brad Tolinski's *Light & Shade: Conversations with Jimmy Page* sheds serious light on this poorly understood, enigmatic musical genius."
— BOSTON GLOBE

"A highly engrossing story. *Light & Shade* is simply an excellent rock 'n' roll read."
— NEW YORK JOURNAL OF BOOKS

"A vivid, immensely interesting and enlightening yet candid history of a genius and a giant. *Light & Shade* is by far one of the best books on the subject related to Led Zeppelin and a portrait of a figure of immeasurable musical and cultural importance."
— ALLABOUTJAZZ.COM

"This is the most comprehensive and compelling collection of interviews, insights, and historical anecdotes of one of rock and roll's premier guitarists, songwriters, and producers ever compiled. A fascinating must-have for Jimmy Page fans like myself."
— SLASH

"*Light & Shade* illuminates the haunted genius of Jimmy Page in an original and completely satisfying way. The conversational dynamic between the author and the subject reveals a wealth of info about the man, the music, and the magick."
— KIRK HAMMETT, METALLICA

"Jimmy Page . . . the one and only! From mild to wild, Jimmy sez it all. This fine work will rock you!"
— BILLY GIBBONS, ZZ TOP

LIGHT & SHADE

LIGHT
&
SHADE

CONVERSATIONS WITH JIMMY PAGE

Brad Tolinski

B\D\W\Y
BROADWAY BOOKS
New York

All rights reserved.
Published in the United States by Broadway Books,
an imprint of the Crown Publishing Group,
a division of Random House, Inc., New York.
www.crownpublishing.com

BROADWAY BOOKS and its logo, B \ D \ W \ Y, are trademarks of Random House, Inc.

Originally published in hardcover in the United States by Crown Publishers, an imprint
of the Crown Publishing Group, a division of Random House, Inc., New York, in 2012.

Library of Congress Cataloging-in-Publication Data
Light and shade: conversations with Jimmy Page/Brad Tolinski. —1st ed.
p. cm.
Includes bibliographical references.
1. Page, Jimmy— Interviews.   2. Rock musicians—England— Interviews.
3. Led Zeppelin (Musical group)   I. Page, Jimmy.   II. Title.
ML419.P37A5 2012
782.42166092— dc23
[B]   2012009454

ISBN 978-0-307-98575-0
eISBN 978-0-307-98573-6

PRINTED IN THE UNITED STATES OF AMERICA

Book design by Alexis Cook
Cover design: Christopher Brand
Cover photography: Redferns

First Paperback Edition

For Kane and Nico Tolinski

CONTENTS

LIGHT & SHADE

OVERTURE

FOR MORE than fifty years, guitarist/composer/producer Jimmy Page has influenced contemporary music in both large and subtle ways. While still in his teens, he and a small handful of musicians helped introduce American blues to the British Isles, sparking a revolution that set the stage for artists such as the Rolling Stones, Jimi Hendrix, and Cream. His virtuoso guitar work on countless studio sessions in the sixties, with artists as disparate as Nico, Joe Cocker, Donovan, and Them, helped create the soundtrack for Swinging London's much-celebrated youth explosion. And his groundbreaking playing, writing, and production work with Led Zeppelin dominated the seventies and continues to resonate decades after.

Even now, Page remains a dynamic force whose inventiveness continues to surprise. His recent "photographic autobiography," *Jimmy Page by Jimmy Page*, is an original and beautifully designed assessment of his life and career, while his new website, jimmypage.com, with its smart graphics and informative content, should satisfy the cravings of his worldwide legion of fans to know what he's up to.

Considering his many accomplishments and rich history, one would assume that there would be several books about him. The world of Jimmy Page, however, has largely been uncharted.

Sounds like another Led Zeppelin mystery, but the truth is that this one is fairly easy to solve. There is the matter of Page's natural reticence; he is, after all, the man who chose to dress himself as a hermit in the band's 1976 concert film, *The Song Remains the Same.* More significant, however, is the fact that Page has had an uneasy, sometimes antagonistic relationship with music journalists and critics—the very same people who tend to write rock and roll biographies.

So why the hostility? Absurd though it may sound, in the early seventies, when Led Zeppelin was coming into its own, the hipster rock press often was, to put it mildly, less than enthusiastic about the band and its now-universally-hailed music.

Rolling Stone magazine was particularly savage. In 1968, critic John Mendelsohn wrote a 389-word vivisection asserting that Led Zeppelin's first album offered "little that its twin, the Jeff Beck Group, didn't say as well or better three months ago." Several months later, *Rolling Stone* chose this same Mendelsohn to review *Zeppelin II*, which he dismissed as "one especially heavy song extended over the space of two whole sides."

Stone was by no means alone in picking on Led Zeppelin. In December 1970, Detroit's legendary rock and roll magazine *Creem* printed a notorious "anti-review" of *Led Zeppelin III*, in which critic Alexander Icenine used faux, drug-addled gibberish to express his utter contempt for the album:

> What is a Led Zeppily? I have oftimes asked of my own selfhead this
> questlung upon retiring to my bed patterns. Or sometimes, how is
> a Red Zipper not a Load Zoppinsky? Many times there is no answer
> and they refuse to do it for ya.

How did Jimmy Page respond to these and other "sober" assessments of his work? He turned his back on the entire rock writing community.

As the band got bigger, the reviews got better and Page's chilly attitude toward the press began to thaw. But in many ways the damage had already been done. Veteran rock writer Jaan Uhelszki recalls one exchange with Page during Zeppelin's 1977 tour that is as telling as it is funny.

"I'd been on the road with the band for over a week and couldn't get Jimmy to do an interview," Uhelszki says. "Finally, on the last day of the tour, he agreed to an audience on the condition that the publicist had to be there. I didn't find out until the time of the interview that Jimmy stipulated that I must first ask the publicist my question and then she would relay the question to him—even though we all spoke the same language and I was sitting a mere six feet from him. This went on for about an hour."

But maybe Page had a right to keep his guard up. Most writers just wanted to know about his alleged drug use, weird groupie sex, or whether it was true that he'd made a pact with Satan. Truth is, few journalists treated him or his band with the respect they accorded his peers John Lennon, Keith Richards, and Pete Townshend. In the long run, none of it really mattered. Jimmy turned his obsession with privacy into an essential part of his mystique. He became rock's greatest enigma.

This is where I come in.

I HAD MY first conversation with Page in the spring of 1993. As editor in chief of *Guitar World* magazine, I assigned myself the job of interviewing Page for a story on his then-recent, and controversial, collaboration with Whitesnake's David Coverdale. But my real interest was far more personal. As a child of the seventies, I grew up with Page's work with the Yardbirds and Zeppelin deeply embedded in my DNA. I had always admired his innovations as a guitarist, composer, and arranger. As a producer, I believe, he ranked up there with true innovators like Phil Spector and George Martin.

As a journalist, I always wondered why nobody ever asked him about *that* stuff, and I imagine Jimmy wondered the same thing. This is what I wanted to read about and wanted to write about.

Page's prickly reputation with journalists was, of course, well known to me, so I prepared myself for a potentially difficult time. I won't say we got on like two bustles in a hedgerow, but I could tell he enjoyed the fact that I was able to converse about his music in a relatively sophisticated and technically knowledgeable way. A couple of hours into our first interview, we hit a small speed bump when he began to humorously feign exhaustion at the perhaps too forensic nature of my questioning. Undeterred, I pushed onward and, miraculously, he hung in there for another hour with absolutely no hint of rock-legend attitude. You could sense that he was happy just to have a serious discussion about his music—not just Led Zeppelin but also his partnership with Coverdale, which had occupied him for more than a year.

Which brings me to the purpose of *Light & Shade: Conversations with Jimmy Page*. In many ways, this book is a natural extension of that first encounter. It is my belief that Page is one of the most important and underappreciated musicians of the last century. I place him without reservation up there with groundbreaking artists such as Muddy Waters, Miles Davis, and Chuck Berry, visionaries who bridged the gap between artistic and commercial success. His music has stood the test of time and continues to intrigue generations of music fans who were born years after Led Zeppelin called it quits. His words, ideas, and story are of historical import.

I made it my goal to induce this famously private man to speak in great detail about his long, storied career as often as I possibly could. And as it turned out, he did so fairly often. Thanks to my position at *Guitar World*, I had many opportunities to chat with Jimmy over the past two decades. While I wouldn't call us friends, our relationship is a friendly one, and we've been able to build a bridge based on our mutual professionalism, which spawned respect.

By "mutual professionalism," I mean he has been forthcoming, polite, and respectful with me as long as I followed a few unspoken ground rules. He expected me to have assiduously done my homework, to have the facts at hand, and to ensure that, to the greatest possible extent, our conversation focused on music. As long as I kept my end of the (again, unspoken) bargain, he was totally cool and answered most questions as honestly as he could.

Here's what he didn't like: open-ended questions, questions that required him to speculate about others' opinions of his music, or anything that required him to say something negative about another artist. Any one of these could cause a perfectly good chat to end abruptly and not resume for a very long time. As a writer, I did occasionally find these restrictions a bit constraining, but it never was a serious problem because there was so much fertile musical ground to cover.

While we're on the subject of the forbidden, it's worth taking a moment to discuss Page's interest in the occult (which is one of the first things people ask me about when they learn that I've spent time with him). Contrary to popular opinion, he has never really hidden his fascination with magick (as British occultist Aleister Crowley spelled it to differentiate the occult from stage magic) and metaphysics, particularly its manifestation in his music. But ultimately he felt that there was very little upside to discussing these subjects in great depth when his comments would only be sensationalized, misunderstood, or taken out of context. He believed it would trivialize ideas that were important to him and make him appear eccentric. Fair enough.

However, it did strike me that his studies are an important component of his art, so I made it a point to fill in some of the missing blanks in regard to his interests when relevant. If this should open a door for those seeking detailed information about ceremonial magick, metaphysics, and astrology, so much the better.

BUT ABRACADABRA ASIDE, *Light & Shade: Conversations with Jimmy Page* has a very specific agenda. It is not a tell-all biography in the traditional sense but rather (I hope) an enlightening and definitive look at the musical life of a rock and roll genius, told in his own words. In the music documentary *It Might Get Loud*, Jimmy briefly touches on what the term "light and shade" means to him: "Dynamics . . . whisper to thunder; sounds that invite you in and intoxicate. The thing that fascinates me about the guitar is that no one ever approaches it the same way. Everyone plays differently and their personality always comes through." Think of this book as an expansion of those basic ideas and a rare opportunity to actually hear a master artist explain his music.

You'll notice that voices other than Page's appear in the book, albeit sparingly. Some were included to add a valuable outsider's perspective on certain historical or musical points made by Page. I also used these voices to add interesting details to Page's narrative. For example, John Varvatos's discussion about Page and his impact on fashion was added simply because I felt he was particularly qualified to address a topic that I believe is an important element of Page's legacy.

Every one of these interpolations was meant to add what Jimmy would call "light and shade" to a picture of a very complex man.

—Brad Tolinski

JIMMY PAGE DISCOVERS THE GUITAR, BECOMES
A LOCAL LEGEND, GOES TO ART SCHOOL, AND HELPS
USHER IN THE BRITISH BLUES BOOM.

Page with his first electric guitar, 1958 (© Tony Busson).

"THERE WAS A FIGHT ALMOST
EVERY TIME WE PERFORMED . . ."

———————

I T IS AN OLD, old story, the heartbeat of many an ancient myth. A young man of humble background stumbles upon a mysterious talisman, the mastery of which would change the course of his life. The boy embarks on a lengthy journey, during which his skills, strength, and mettle are tested to determine his worth. He ultimately unlocks the awesome power of the talisman, leading to great glory for himself and, often as not, a reordering of the cosmos. So it was with Jimmy Page, founder of Led Zeppelin and one of rock's greatest guitar legends.

On Sunday, January 9, 1944, James Patrick Page was born to parents James and Patricia in the London borough of Hounslow. The family stayed in the area for nearly a decade, until the noise from nearby Heathrow Airport prompted them to move to the quiet suburb of Epsom, in Surrey. Or, as Page dryly remarks, "When the jets arrived, the family left." It's here that the real story begins.

"The weirdest thing about moving to Epsom was that there was a guitar in the house," Page told British journalist Charles Shaar Murray in 2004. "I don't know whether it was left behind by the people before, or whether it [belonged to] a friend of the family's—nobody seems to know how it got there."

It would be a stretch to suggest that Jimmy's discovery of a mysteriously discarded guitar was an act of divine providence. However, it is indisputable that a man whom millions would one day call the King led Page to realize that his destiny was linked to his mastery of that gift guitar: Jimmy has stated that Elvis Presley's recording of "Baby Let's Play House," featuring the slicing, reverb-soaked rockabilly licks of guitarist Scotty Moore, was one of many key tracks that inspired him to get serious about music. "I heard that record and I wanted to be a part of it," he explains. "I knew something was going on."

At the age of thirteen, Page learned to tune his guitar from school friends and to strum some rudimentary chords from some local players, but beyond that he was largely self-taught. He learned by ear how to play songs from recordings by British skiffle sensation Lonnie Donegan and early American rockers like Presley, Eddie Cochran, and Gene Vincent.

Impressed by his dedication, Jimmy's dad bought him an acoustic sunburst Hofner President f-hole guitar, which resembled the big Gibson guitars played by his heroes Moore and Chuck Berry. In little over a year, Page was already good enough to perform two songs on the BBC-TV program *All Your Own*, a talent show for teens hosted by Huw Wheldon. Videos of the 1958 performance show the precocious Page bopping with confident enthusiasm while playing the novelty song "Mama Don't Allow No Skiffle Around Here" and Leadbelly's "Cotton Fields."

Soon afterward, Page bought his first solid-body electric guitar, a three-pickup 1958 Resonet Grazioso Futurama, which resembled the sleek Fender Stratocaster guitars favored by rock stars such as Buddy Holly. He continued to hone his craft while playing with a series of local Epsom bands, and in 1960 he caught the eye of music manager Chris Tidmarsh, who sought to recruit him for a gang of rockers called Red E. Lewis and the Red Caps, whose very name was a tribute to Jimmy's heroes, Gene Vincent and the Blue Caps.

Page remembers those early shows as being fun but rowdy. "I was still in school, so we would only play on weekends," he says. "But it was an eye-opening experience. There was a fight almost every time we performed. It wasn't like the fights you have these days, where people get shot, stabbed, or killed. It was more like violent sport. Basically, the first guy to hit the floor lost, and that would be it. But I had to learn how to keep my head down and play through all kinds of situations."

Several months after Jimmy joined the Red Caps, Tidmarsh, who changed his name to Neil Christian, fired Red E. Lewis and made himself the band's singer. He renamed the band Neil Christian and the Crusaders and aggressively hit the road, playing up and down the English club circuit.

A big part of the group's popularity was the boy wonder Page, who was able to replicate the popular sounds of the day with his newly acquired orange Gretsch Chet Atkins Country Gentleman. From the high-energy rock and R&B of Chuck Berry and Little Richard to slower instrumentals like Santo and Johnny's "Sleep Walk" to whatever was in the Top 20, Jimmy could play it all, and do so with flair.

While the shows were always exciting, the living conditions, pace of the performances, and tough travel itineraries were emotionally and physically punishing. For the next two years, Neil Christian and the Crusaders lived out of the back of their van and in the clubs they headlined at, sleeping on floors or on top of their instruments.

One night in the summer of 1962, Page collapsed after a gig. He was diagnosed with a form of mononucleosis, and soon thereafter he gave his notice.

JIMMY'S INTRODUCTION TO the entertainment business had been rough-and-tumble, but there was no doubt that, by the age of eighteen, he had become a polished guitarist, mature beyond his years. His reputation had grown to such an extent that, even while he was in the Crusaders, he had been asked to play on a 1962 recording session with two of England's most respected rock musicians: bassist Jet Harris and drummer Tony Meehan, both of whom played with one of Britain's biggest bands, the Shadows. The song they recorded was "Diamonds," an instrumental composed by Jerry Lordan, and it became a number-one smash in the UK charts in early 1963. It was also during this period that UK blues harmonica virtuoso Cyril Davies approached Page to join his influential R&B All-Star Band.

But after his experiences with the Crusaders, Jimmy was wary about becoming a touring musician.

While recovering from his illness, Page began to consider his prospects. He loved playing guitar, but his time in Neil Christian's band gave him second thoughts about making music his career. He had been doing a lot of painting and drawing in his free time and decided to take a prediploma course at Surrey's Sutton Art College. For the next year and a half, Jimmy diligently pursued his formal studies, but perhaps just as diligently, he continued to play the guitar.

Page began spending his evenings haunting the small but growing London blues-club scene, jamming at places like London's Marquee Club and Richmond's Crawdaddy Club. The British blues boom was in its embryonic stages, but he was already well versed in the music of the American South. His interest had been piqued years earlier by his beloved rockabilly, but local R&B buffs and record collectors fanned the flames.

Just as he had devoured the licks of Gene Vincent guitarist Cliff Gallup and Ricky Nelson guitarist James Burton, Page greedily consumed the solo and rhythm styles of blues players like Hubert Sumlin, Elmore James, and Memphis Slim guitarist Matt "Guitar" Murphy. During his time in the Crusaders, Jimmy attempted to incorporate his new passion into the band's repertoire, but the music did not sit well with the mainstream ballroom-dancing crowds that constituted its main audience.

The times were changing, however, and a year or two later, British music fans began taking greater interest in black American sounds. In the north of England, the Beatles were having great success playing songs from Motown's dance-oriented R&B catalog. But in the south, a small and devoted group of musicians took to studying and performing the raw electric blues released by Chess Records and other Chicago-based labels. In 1962, harpist Cyril Davies and guitarist Alexis Korner opened a new Thursday residence at the Marquee Club with their band Blues Incorporated. The gigs became a meeting place for hipsters and musicians who enthusiastically embraced the music of Muddy Waters, Howlin' Wolf, and other Chicago blues greats.

Soon, Jimmy was invited to lead a band that performed at the Marquee during the interval between the headliner's sets.

During that period, Page realized the extent to which he was still passionate about the guitar. While he pondered his future, fate intervened when he was invited to play on several more recording sessions. Soon he began to think that a career as a studio guitarist might be a good way to earn a living without having to tour.

CONVERSATION

Q:

YOU STARTED PLAYING THE ELECTRIC
GUITAR WHEN IT WAS STILL A
RELATIVELY EXOTIC AND UNUSUAL
INSTRUMENT. WHAT INSPIRED
YOU TO PICK IT UP?

Like many young people of the era, I loved the guitar-driven rockabilly of Elvis and Gene Vincent. It's amazing to me now: The guitar parts were so subdued, but I was so engrossed that they seemed very loud—right up there. I just used to listen to my music, and in my mind, I would go back through the cone of the speaker into a world of my own. I would pretend that I was sitting in the studio with these artists and engineers and we'd study the echo and how the music was created. I might've been deluding myself, but I thought I could tell the difference between the recorded sound of one particular session from another and what was being applied. Certain echoes and reverbs seemed earth-shattering. Now when I listen to those same records, all of those effects are way in the background, but that's how hard we studied these records, and that's how hungry we were. All of us—Eric Clapton, Jeff Beck, and our contemporaries—went through the same process. Those early rock and blues records grabbed us hard.

When did the blues come into the picture?
It didn't take me long to notice that some of my favorite Elvis songs, like "Hound Dog" and "Milk Cow Blues," were originally written and recorded by blues performers. We began to discover people like Arthur Crudup, who wrote Presley's hit "That's All Right." So in this way, bit by bit, you start understanding a much bigger musical picture. You discover that music is a tapestry that unfolds.

I started going back to the source of his music through a friend of mine that was a record collector. He had an amazing stash of blues albums, and he was very generous about letting me listen to them. No one was really playing the blues on the radio or in clubs yet, so it was still an underground thing; records were very hard to find.

It's not hard to see why I gravitated to rock and blues. I was a guitarist and it was a very guitar-centric music. If you were a guitarist at that time, your appetite was voracious for Chuck Berry and all the blues that was coming out of Chicago.

The fact that the blues dealt with sex and the devil must have also made them attractive to a young guy.

When I heard those songs for the first time, they really did send chills up my spine. They still do.

What saved the day was that there were other people that just really loved rock, blues, and R&B, and they also began collecting these obscure records. Soon a whole network formed of people who would swap and trade music. They'd lend you a record so you could work out certain solos. None of us really had any money to buy all of these rare imported albums, but it all built up. It was a very, very important period.

Besides record collectors, who were some of the other heroes who brought the blues and rock to England?

Well, you would have to mention Alexis Korner and Cyril Davies, who had a band called Blues Incorporated. Alexis played acoustic guitar and Cyril was an amazing electric harp player, and back in the early sixties they would host these regular blues jams on Thursday nights at the Marquee Club. It was the only thing like that in London at the time. The Rolling Stones played there before they became famous, Clapton would be in the audience, and I would regularly participate in jam sessions that would happen between sets.

Alexis also brought blues artists like Muddy Waters and Sonny Boy Williamson to England for the first time, which was incredible.

Was the Marquee a big place?

I guess it held a couple hundred people. It seemed very big at the time to me, I'll tell you that! [*laughs*] It was a big gig for me.

I remember one night Matthew Murphy came to play the Marquee. The place was packed, because we all loved his playing. We were all psyched and ready for him to rock out, and he looked at us and said, "Naw, man, I just want to play some jazz." Everybody just groaned. It was very funny.

What kind of music were you playing at that time?

I was trying to play like Matt Murphy! [*laughs*] I think I was also playing some Freddie King.

Was the release of Robert Johnson's *King of the Delta Blues Singers* in 1961 a significant event in the UK?
It was significant, but it took a little while to get around the grapevine. But, believe me, there was a grapevine. That's how we heard about Freddie and Albert King, Robert Johnson, and a number of other country bluesmen.

What did you think of your white, blues-playing contemporaries at that time? Did you like the Stones, the Animals, and the early Yardbirds, or did you think they were jive for trying to play black music?
There was no real snobbery; we were all trying to do our own take on the blues at that time. I had heard about the Stones from the recording engineer Glyn Johns. I was working as a session musician at the time, and he would rave about them. I finally went to see them, and I was really impressed. They really had the Muddy Waters groove dead-on. Brian Jones in particular was playing very authentically.

A milestone in the development of British blues was when Sonny Boy Williamson actually played and recorded with Eric Clapton and the Yardbirds. Clapton, who was in the band at the time, has said it was a real education for him, but he didn't think it turned out well. Again, what did you think of their collaboration?
When I first heard about it, I thought it was really exciting. I mean, no one really expected the Yardbirds to sound like a Chess band, but I thought they did a really credible job. They had their own take on the blues, and there's nothing wrong with that. There was just so much going on at that time, and everyone was just trying to push the music further.

So you saw the British movement as just another step in the evolution of the blues.
Something like that.

Is it fair to say that Led Zeppelin used the blues as a vehicle to create something modern—almost futuristic?
Sure. Well, we had this magic rhythm section, didn't we? It took whatever

we were playing into another dimension. It allowed Robert Plant and I to really stretch out. But we weren't the only ones—you have to remember that Cream and Jimi Hendrix were also exploring similar territory. Hendrix took the blues into outer space, didn't he?

It seems that there were two camps when you guys were coming up. There were the purists who were trying to completely capture the sound and spirit of Chicago blues. And there were those who were more interested in trying to stretch the form. I would say you and Jeff Beck belonged to the second group. Were you considered heretics?

Jeff and I had a broader view of music, but I never saw it as a battle with more traditional blues players. I admired what purists like Clapton were doing, and there were plenty of others who were equally brilliant. Personally, I don't think you're going to find a better example of British blues than the original Fleetwood Mac, with Jeremy Spencer and Peter Green.

While you first became prominent as a member of the Yardbirds, you had an interesting musical career before that. You played with Neil Christian and the Crusaders and then went on to become a top session musician and producer. Could you give us a little sense of your life as a guitarist in the very early sixties?

I was just a teenager when I played with Neil. We played a bit of everything, really—Chuck Berry, a bit of Chet Atkins, and a lot of the pop music of the day. Yeah, it was the days of the orange Gretsch.

We acquired a good reputation, but touring was very primitive and I found it very difficult at the time. For example, I remember we were driving to a Liverpool club once and the van broke down and we had to hitchhike. By the time we arrived, we were so late that we only had time to play for forty-five minutes. And because we had to hitchhike we had only guitars, so we were forced to play through other people's amps, which sounded terrible. We didn't really have any money, so we ended up sleeping in this little room in the club, in the middle of the desk chairs and the fucking first aid cabinet, and it was really cold.

Anyway, because of all the traveling and adverse conditions, I kept getting glandular fever. After a while, I thought, the hell with it.

I decided to pack it in and go back to art college, which I really enjoyed. At the same time, I was only eighteen and hadn't really made my mind up about what I was going to do with my life.

How did you get back into music?

I never really stopped. For example, I'd have these Sunday-night jam sessions at my parents' house with Jeff and other musicians. Then, in the middle of '62, Alexis Korner and Cyril Davies began organizing these blues jams at [London's] Marquee Club on Thursday nights. These became quite popular. In fact, you could trace the whole beginning of the British blues boom to those jams. All the musicians would flock there, because it was the only game in town. I started regularly joining them onstage, which is where I met Clapton and members of the Stones. This was before any of us had any real notoriety.

During that period I realized that I was still really passionate about playing. As the Marquee shows started becoming more popular, I started getting approached to play studio sessions.

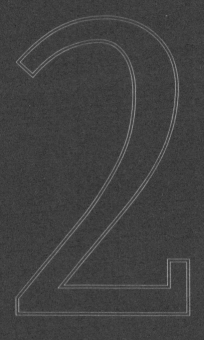

JIMMY BECOMES ONE OF BRITAIN'S BUSIEST
SESSION GUITARISTS AND THEN THROWS IT
ALL AWAY TO JOIN THE YARDBIRDS.

Page with Carter-Lewis and the Southerners, 1963 (© *Getty Images*)

"I WANTED TO PLAY LOUD . . ."

PAGE'S CAREER AS a studio guitarist started in early 1963, but it gained serious momentum later that year when he was asked to play on a session with John Carter, a singer with a studio-based group called Carter-Lewis and the Southerners. Carter's "Your Momma's Out of Town," featuring Jimmy's guitar work, became only a moderate hit, but that was enough to fuel his reputation as "the new pick on the block."

Legendary producer Shel Talmy was particularly impressed by the young guitarist's command of rock and blues, and he asked Page to contribute to the Kinks' 1964 debut album and the Who's landmark first single, "I Can't Explain." But that was just the tip of the iceberg. Jimmy can be heard on literally hundreds of tracks from that era, including Tom Jones's smash hit "It's Not Unusual," Shirley Bassey's dramatic "Goldfinger," and the intense garage rock–blues classic "Baby Please Don't Go," by Them. He was so in demand, it's estimated that his guitar can be heard on 60 percent of the records recorded in Britain during the early sixties.

During this time, Jimmy relied primarily on the black Les Paul Custom he bought while he was in Neil Christian's band. With its three humbucking pickups, the Fretless Wonder, as it was nicknamed, afforded Page wide tonal flexibility, making the instrument perfect for just about anything he was called on to play. For acoustic-oriented work, he played a 1937 Cromwell archtop guitar. His amplifier of choice in those days was a Burns, although in 2009 he recalled occasionally using a Fender.

Electronic effects had only just begun to appear on the market at that time, but Page, by now a mover and shaker in the very epicenter of the London music scene, was often first to learn about the latest technological innovations. For example, when a then-unknown inventor named Roger Mayer developed one of the first truly usable fuzz pedals, Page, recognizing its possibilities, immediately put it to use.

"The first fuzz box I made in 1964 was the one Jimmy used on some of the P. J. Proby records," Mayer told journalist Mick Taylor. "He was doing two or three sessions a day back then, and as a result, quite a few hit records had some of my fuzzes on them." Mayer went on to develop a whole array of revolutionary pedals for the likes of Jeff Beck, Dick Taylor of the Pretty Things, and Jimi Hendrix, who used one of Mayer's distortion units on "Purple Haze."

After serving two years in the studio trenches, Jimmy was at the top of his game. In an incredibly short period, he had become proficient enough in a broad array of styles in multiple genres to deliver dynamic solo and rhythm parts under great pressure. The young musician also kept his eyes open and learned how to produce studio sessions by observing England's best producers at work. Soon he began to construct his own theories on recording, mixing, and engineering, and how rock records should be made.

But Page's musical education didn't stop there. Even outside of the studio, he was ever alert to new and esoteric musical trends. He was one of the first musicians in England to acquire a sitar—years before the Beatles' George Harrison did—and learned how to tune the instrument from the legendary Ravi Shankar. He also kept close tabs on the burgeoning British traditional-folk-music scene percolating in London's coffeehouses. Acoustic virtuosos such as Davey Graham, John Renbourn, and especially Bert Jansch had a profound effect on Page, introducing him to the alternate guitar tunings and finger-style techniques he made his own on future Zeppelin classics such as "Black Mountain Side" and "Bron-Yr-Aur."

Everything Page learned at the time would serve him well down the line, but in the meantime he found that the novelty of life in the studio was beginning to fade. As the sixties galloped on, bands became more proficient on their instruments. As a result, Page was playing fewer exciting rock sessions and getting more commercials and jingles. For every stimulating studio date—such as the one in which he assisted the Rolling Stones' Brian Jones on a soundtrack for German director Volker Schlöndorff's film *A Degree of Murder*—there were a multitude of mind-numbing sessions for Muzak-oriented projects. Pap, rather than pop, dominated his days, and Page's once-idyllic recording-studio gig began to resemble a lifetime sentence in a windowless cell.

In 1965, about a year before he became completely disenchanted with his session work, Page received an offer from the Yardbirds' manager, Giorgio Gomelsky, to replace Eric Clapton, who was leaving the pioneering blues-rock group. The Yardbirds had just scored their first number-one hit, "For Your Love." Page, however, was wary of offending his friend Clapton, so he recommended his childhood pal Jeff Beck for the gig.

While he had no regrets about handing over the plum job to Beck, Jimmy, tired of his unofficial role as "the hidden face behind the hip hits," had already taken some subtle—and not so subtle—steps to free himself from his studio shackles. In early 1965, he released a delightfully aggressive solo single on the Fontana label entitled "She Just Satisfies," on which he sang and played all the instruments except drums. The track featured a great pile-driving guitar riff, snarling lead vocals, and a surprisingly strong harmonica break.

Later that year, Page received another compelling offer. Rolling Stones manager Andrew Loog Oldham and his partner, Tony Calder, asked him to become a staff producer for their successful independent label, Immediate, home to sixties luminaries such as the Small Faces, the Nice, and Fleetwood Mac. The guitarist jumped at the chance. Among his assignments,

Page produced a single for German chanteuse Nico, who would soon move to New York City and work with the Velvet Underground.

More significantly, Page produced several groundbreaking 1965 recordings for John Mayall and the Bluesbreakers featuring ex-Yardbird Eric Clapton. Among the tracks he made for Immediate were the remarkable "I'm Your Witchdoctor" and "Telephone Blues." Page, impressed with Clapton's big, muscular guitar sound, was determined to capture every nuance on tape.

Page recalls, "Eric was using a Les Paul with one of Jim Marshall's new amplifiers, and it was a perfect match. It was a funny session, though, because Eric was using feedback to sustain certain notes, and the engineer had never heard a guitarist use that technique, so he kept pulling down the faders. The guy couldn't believe that someone was getting that kind of sound out of a guitar on purpose."

The significance of this session cannot be emphasized enough, for it represented the birth of the modern guitar sound. And while Clapton did the playing, it was Page who made it possible for his work to be captured properly on tape.

CONVERSATION

Q:

WHAT WAS THE STUDIO ENVIRONMENT
LIKE WHEN YOU STARTED
TO PLAY SESSIONS?

I was still very young—most of the guys doing sessions were twice my age—but they needed a kid from the streets to play on the rock-oriented dates. But soon I started playing on all sorts of things, including acoustic guitar on folk albums and rhythm on jazz sessions. It's quite terrifying now, when I think of all the things I did, but I took it on. I took the bit in my teeth and went for it. It was a great apprenticeship.

Did you miss playing in a band?

Not at first, because as far as I was concerned, being a studio musician was cool. When I first started playing sessions, I didn't know how to read music, but ultimately I had to because the work became more complicated. The first sessions I played, the producer would just say, "Oh, play what you want. Great." But soon that wasn't the case. The more sessions I played, the more they wanted you to just find the crows on the telegraph wires—you know, just play what was written. Eventually it became boring.

While you were playing sessions, you still seemed to be keeping your finger in the outside world. Didn't you procure some studio work for Jeff, and didn't you produce a couple of pretty solid post-Yardbirds Clapton sessions?

Yes. At one point I was contracted to work as a staff producer for Immediate Records, and I used Jeff on a few sessions. It was a similar situation with Eric. I was approached by Immediate Records to produce some tracks for a special British blues series [*An Anthology of British Blues, Vol. 1* and *2*]. We cut four songs with Eric, who was just starting to work with John Mayall, including "Telephone Blues," which I think has one of his best solos.

The Yardbirds asked you to join after Eric left.

Yeah, they actually asked me twice. They asked me before Eric had even left the band, because their manager, Giorgio Gomelsky, wanted to get more commercial and Eric wanted to get more purist; Giorgio wanted to force him out. Then they asked me again when Eric finally left. But I was still too nervous about getting ill while on the road, and I wasn't quite sure about the politics with Eric, because we were friendly. So I recommended Jeff Beck, who I think was amazing as far as pushing the Yardbirds to the next level. His imagination on those Yardbirds albums is incredible.

Unfortunately, soon after that, I started getting a lot of sessions that just weren't fun. So when Jeff approached me later to join him in the Yardbirds, I was anxious to do it. Still, I don't regret the studio work, because it was excellent training.

You were experimenting with echo, distortion, and feedback back in '63 and '64. However, you never received much credit for that, because you did it behind the scenes, on your many recording sessions. What do you recall about the development and use of guitar effects in those early days?
A turning point came when Roger Mayer began making his distortion boxes. I remember playing this gig in the early sixties when Roger came up to me and said he worked at the British Admiralty in the experimental department, adding that he could probably build any electronic gadget that I wanted. I suggested that he should try to make something that would improve upon the distortion heard on "The 2000 Pound Bee" by the Ventures. He went away and came up with the first real good fuzz box. It was so great because it was the first thing that really generated this wonderful sustain. After that, he swept through the British music scene. He made one for Jeff, one for the guitarist in the Pretty Things, and then he started working for Jimi Hendrix.

There seems to be a real quantum leap in recording from, say, the early Beatles and Yardbirds records of '66 to Led Zeppelin's debut in early '69. A lot of the earlier records lacked the sonic depth of music recorded in the late sixties.
I don't know if I completely agree with that. The stuff recorded at Sun [Studio, Memphis] still sounds great in every way, but I think what you may be referring to are a lot of the early albums that came out of England. You have to remember, a lot of early-sixties bands were subjected to producers and engineers that didn't really understand or like rock music. The quality of records started to change when the bands started taking more control, and more sympathetic engineers like Glyn Johns and Eddie Kramer began getting involved.

The first Zeppelin record, for example, probably sounded so good because I had so much experience in the recording studio and I knew exactly what I wanted and how to get it. That's a tremendous advantage. Also, the quality of recording in general improved with the introduction of the eight-track [reel-to-reel] machine. That just opened up a world of possibilities.

Your contribution to the art of recording has been significant. The first Zeppelin album set a new standard for depth and dynamics, and you played an integral role in creating the sound.

Essentially it came down to mike placement—moving the mike away from instruments in order to give the sound a chance to breathe. In my session days, I worked with this amazing drummer named Bobby Graham. You'd see him set up in this little recording booth, with a mike shoved right next to his kit, and he'd be whacking the hell out of the drums, yet the recorded sound would be tiny.

It didn't take long to figure out the reason why. Drums are an acoustic instrument, and acoustics need to breathe. It was as simple as that. So when I recorded Zeppelin, particularly John Bonham, I simply moved the mikes away to get some ambient sound. I wasn't the first person to come across that concept, but I certainly made a big point of making it work for us.

In a weird way, your approach to miking represented a return to the style of Sam Phillips, whose use of ambient miking created Sun's signature sound.

It's true. You want to capture the expansion of that sound, even if it soaks up great, huge chunks of space. That's what made rockabilly sound great.

Who were the greatest drummers that you've heard or played with?

It's sort of obvious, but I've got to say John Bonham. I don't know if you've heard any of the bootleg tapes of, say, "Trampled Underfoot," when he's trying all these other rhythms and he's moving around so much on the ride cymbal. His independence was just second to none. All you ever hear about jazz drummers is their independence. He just pissed all over them.

You produced Jeff Beck's landmark 1967 instrumental "Beck's Bolero," featuring Jeff on guitar, Keith Moon on drums, John Paul Jones on bass, and Nicky Hopkins on piano.

Yeah, and I played rhythm guitar on a Fender electric 12-string. I have to admit, it was pretty memorable. Moon smashed a two-hundred-and-fifty-dollar microphone while we were recording by just accidentally hitting it with his stick. Halfway through "Bolero" you can hear him scream, then hit the mike, and from there all you hear are the cymbals. The song just continues. It was sort of funny.

Before you formed Led Zeppelin you were the lead guitarist in the Yardbirds, and before that you were one of England's leading session guitarists and an upcoming producer. What impact did those experiences have on your work with Zeppelin?

They were very valuable. I learned an incredible amount of discipline. I learned how to read charts and started playing on things you'd never expect, like film scores and jingles. I even played some jazz, which was never my forte. But having to vamp behind people like Tubby Hayes, who was a big jazz saxophonist in England, or play on several of Burt Bacharach's pop sessions gave me a fantastic vision and insight into chords. Being a session player, however, wasn't really me—it wasn't rock and roll. I wanted to play loud!

MUSICAL INTERLUDE

A CONVERSATION WITH
JIMMY PAGE AND JEFF BECK

AS TEENAGERS, JIMMY PAGE AND JEFF BECK JAMMED
AND TRADED LICKS. AS YOUNG MEN, THEY CHANGED THE
SOUND OF ROCK AND ROLL. DECADES LATER, THE SCOPE OF
THEIR INFLUENCE ONLY CONTINUES TO GROW.

JIMMY PAGE, Jeff Beck, and Eric Clapton—perhaps the three most important electric guitarists in rock history—were all raised in towns southwest of London, just miles apart. It's as if Willie Mays, Mickey Mantle, and Hank Aaron belonged to the same Little League.

"There must've been something in the water, mate," Beck says with a laugh.

"It's fascinating, isn't it?" Page says. "What was the radius—twelve miles? What I find particularly striking is that even though we didn't know each other when we were growing up, how similar our stories are. We were the freaks—the one in four hundred kids at our schools that played the electric guitar."

"But eventually we all met up, because we started hearing rumors about the one or two other people who played weird guitars," Beck says. "You know, the light shines from another town."

And shine it did, with a little help from Jeff's older sister, who introduced Beck to Page in the early sixties. "She was going to art school, and she told me about this guy at school who also played a 'funny guitar' like mine," Beck recalls. "I said, 'What kind of funny guitar?' She said, 'Oh, I don't know. It's one of those weird-shaped guitars.' And that was that. I got on a bus and went over to Jimmy's house. I think we were about sixteen or seventeen."

Not surprisingly, the two players struck up a fast friendship. Equally starved for information, they began spending hours jamming together at Jimmy's house, trading licks and sharing the few scraps of information each was able to learn about the rock and blues heroes of the fifties. From that point on, Jimmy and Jeff's paths were permanently intertwined.

And even after forty years of living the rock and roll *vida loca*, it is clear that Beck and Page remain that rarest of things—close friends. Without these two, it would be a very different guitar world indeed. Things we take for granted—distortion, feedback, power chords, extended jamming, false harmonics, exotic tunings, and controlled use of the whammy bar—were all pioneered by Beck and Page. As much as Chuck Berry, Elvis, and the Beatles, these two giants shaped the sound of rock and roll.

Why do you think you both progressed so fast? Was it because you were pushing each other?

JEFF BECK Yeah, I used to be very thrilled that Jimmy was living so near. You need a pal to bounce ideas off.

But my sister was also a very important part of my progress, because she used to bring the records home. She was four years older, so she had some money and could swing abroad and buy the new rockabilly records. And you had to have the albums to learn from, because you would never actually hear rock and roll on British radio. I mean, you might occasionally hear "Be-Bop-a-Lula" by Gene Vincent or "Lucille" by Little Richard, but it'd be on some jive request program that you would have to be glued to for hours on end, just in hopes that they would play something interesting.

JIMMY PAGE There was one point where British radio wouldn't play rock and roll at all—it had become a dirty word. So then we had to try and tune in the American Forces Network out of Germany.

Do you remember any of the specific licks that you'd show each other when you were hanging out?

Jeff Beck and Page (© *Ross Halfin*)

BECK We would play Ricky Nelson songs like "My Babe" and "It's Late" because his guitarist, James Burton, was so great. And just a lot of jamming. I remember Jim had a two-track tape recorder, which was a dream. He used to stick the mike, which came with the tape recorder, under a cushion on the couch. I used to bash it, and it would make the best bass drum sound you ever heard!

What kind of guitars did you have in the early days?
PAGE At that time I think I was playing the Gretsch Country Gentleman.
BECK I had a few different guitars in the beginning, including one that I

made myself. I had a Japanese Guyatone, a Burns, and then finally a Fender Telecaster.

How about amps?

BECK I had an amp and a speaker cabinet that I made; it was totally huge and ridiculous, and it kind of took over our whole house. My poor parents! It had this little speaker in it, but it created this huge wall of sound.

Where did you get the know-how to build an amp?

BECK You'd just go to an electronics store and buy a chassis with some tubes, pick some speakers, and then you'd just build a cabinet around it. I bought speakers at my electrical shop and would ask the guy how to get more treble. And the owner said, "What do you mean? This has got all the treble you need. You see, all you do is take bass away." And I thought, That's a cool thing. Just get rid of all the bass.

He also ordered me these speakers called Axiom 300s, which would tear your head off with treble frequencies. I just was a treble freak, and my mum used to go bananas because her ears were not quite attuned, shall we say, to that kind of volume.

In terms of watts, what are we talking about?

BECK We're talking about ten, seven watts. I don't know.

And that was enough to send your parents through the roof?

BECK Forget it! Ten watts in a small room will hurt you if you crank it up.

PAGE I remember playing through my parents' radio. I think that was the first time I went electric. I discovered at some point that my parents' radio had an input on the back. When the sound came through the speakers I couldn't believe it.

BECK Yeah! You could imagine yourself actually being on the radio. That was the coolest thing in the world. You could actually interject your own little solo into a record! The balance was not very good, but it was a magic toy.

What was your first commercial amplifier like?

BECK I couldn't really afford one for a long time. I remember just trying to whip up enough money to go to London with my friend, who was a guitar freak, and together we would just bug the shit out of the people who worked at the guitar store. We'd see these massive amps, like a Fender Bandmaster or something, and we'd think, Oh my God, I've got to have one of those. But the money just wasn't there. So we'd just plug into a quality amp and play until they would chase us away, and that would keep us buzzing for the next six months, just thinking about the great sound it made.

Jeff, did you have any reservations about taking Clapton's place in the Yardbirds?

BECK Not for a second. I was playing in a very good band called the Tridents, and they were always raving about the Yardbirds. I had never really heard them, but they were always talking about Eric Clapton this and Eric Clapton that. I can tell you, I was getting pretty tired of that adulation for someone else. I was like, "Fuck Eric Clapton—you know, *I'm* your guitarist."

And then one day we were in this little store and there was a little transistor radio playing the Yardbirds' version of "Good Morning Little School Girl," with Eric on guitar. And I went, "Oh great, fantastic!" But I didn't really think it was all that impressive.

So I got a little bit of courage and, the next thing you know, I'm in the bloody Yardbirds, facing Eric's audience at the Marquee. I was a little nervous, but I also knew that it was the best break I was ever going to get. So I just went for it. And luckily I had a great night. I pulled out every trick I knew and got a standing ovation.

After that, the big test was to play this club in Richmond, because that's where all the real blues fanatics went. It was kind of an athletic, smelly kind of place to play, and the audience would actually stand on each other's shoulders. It was the first time I really felt like I was going to be slaughtered. But I was cocky. It was like, "All right, you bastards, get a load of this!"

Your wilder, more eclectic style of playing was quite a departure from Eric's. Were the Yardbirds immediately receptive to your approach?

BECK Yeah, they were wonderful. But it might have been because they were just coming off the success of their first big hit, "For Your Love," which was already lifting them out of the club circuit. If they hadn't had that success, they might have looked to me to create some new excitement, which would have been quite a burden.

But before I knew it, we're flying off here and there because the record was doing well, even though I never played on it. We took off like a rocket and I thought it was great. I mean, I had the best job anyone could ever have. The Yardbirds already had a reputation, so I just waltzed in. I didn't even have to buy a new suit to match the rest of the band—I just wore Eric's old one, which fit me just fine.

Yeah, but were the other members of the Yardbirds receptive to your warped playing style?

BECK Well, on the next single, they let me go mad on the B-side. I did "Steeled Blues," and all that, to keep the blues thing going. But then they started pushing me to be more outrageous, like, "Can you bring in some of your trick sounds?" So I started bringing all the gizmos and techniques I used in the Tridents—echo, distortion, hitting the amp, feedback, stuff like that.

Your first major hit with the Yardbirds was "Heart Full of Soul." It was distinctive for two reasons: First, it was one of the first big rock songs to incorporate a distinctly Eastern-flavored riff, and second, it was the first significant hit single built around fuzz-box distortion, preceding by several months songs like the Rolling Stones' "Satisfaction" and the Beatles' "Taxman."

BECK Well, the Yardbirds had just had a hit with "For Your Love," which featured a harpsichord, so they wanted to try using other instruments. So they hired classically trained Indian musicians to play sitar and tabla on "Heart Full of Soul." The only problem was, they couldn't understand four-four timing. When they left, I had the riff going through my head and I just played it playing octaves on the middle G string. Then, by bending the notes slightly off key, I duplicated the sound of a sitar. I then got the idea to use fuzz to dirty up the amp.

Jimmy was actually at the studio at the time, and I borrowed his Roger Mayer fuzz box to work out the idea. Then, when I went to record my part later, I used a Sola Sound Tone Bender, which was one of the first fuzz boxes available commercially.

Jimmy, what's your favorite Jeff Beck performance?
PAGE I still remember the time Jeff came over to my house when he was in the Yardbirds and played me "Shapes of Things." It was just so good—so out there and ahead of its time. And I seem to have that same reaction whenever I hear anything he does.

Jeff, what is your favorite Jimmy Page performance?
BECK Golly, what can I say? I have such an inner sense of pride when I see people waxing so lyrical about Led Zeppelin and to know where it started. There's a much bigger picture there, bigger than selecting something he's done. I'm partial to "Kashmir," but whenever I hear Jimmy on the radio I immediately think of all the great times we've had and the music we've played.

What about Eric Clapton?
BECK He's the ambassador, isn't he? He's the guy that everyone makes reference to. He's the household name for electric guitar.
PAGE He's definitely got the touch, you know. "I Ain't Got You," with the Yardbirds, was really, really great. And his work is still top-notch. The touch is still there.
BECK Eric's got so much to offer. In addition to playing so well, he gives the world songs they can identify with.

JIMMY JOINS THE YARDBIRDS, LEARNS HOW
TO BE A ROCK STAR, AND BUILDS THE
FOUNDATION FOR LED ZEPPELIN.

The Yardbirds: (from left) Jim McCarty, Chris Dreja, Keith Relf, Page, and Jeff Beck, 1966
(© Bowstir Ltd. 2012, Gered Mankowitz/mankowitz.com)

W HILE PAGE WAS growing restless in the studio, the Yardbirds, with Keith Relf on vocals and harmonica, Paul Samwell-Smith on bass, Jim McCarty on drums, Chris Dreja on rhythm guitar, and Beck on lead guitar, began racking up hits like "Heart Full of Soul" and "Shapes of Things," recordings that managed to be both innovative and commercially successful. Unfortunately, the Yardbirds' string of hit singles was purchased at a price: The band had to endure punishing American "package tours" and hasty recording sessions.

Bassist Samwell-Smith tried to improve the situation by seeking more control. He formed an alliance with the Yardbirds' second manager, Simon Napier-Bell, to take over production of the group's records. The bassist acquitted himself well and produced some of the band's best tracks, but even with increased artistic influence, Samwell-Smith continued to grow discontented. When he left the Yardbirds in mid-1966, Jeff Beck was quick to recommend Jimmy Page as a replacement.

"Jimmy wasn't a bass player," Beck says. "But the only way I could get him involved was by insisting that it would be okay for him to take over on bass in order for the band to continue. Gradually—within a week, I think—we were talking about doing dueling guitar leads, and then we switched Dreja onto bass in order to get Jimmy on guitar."

By this point, Page was more eager to join the Yardbirds than he had been a year earlier. While session guitar playing was certainly profitable, the hack element that went with the job was taking its toll. Additionally, the guitarist was ready to step out of the shadows and show the world what he could really do.

The first Yardbirds recording Jimmy and Jeff made together was "Happenings Ten Years Time Ago" in September 1966. A tour de force of guitar invention and orchestration, the track ranks among the greatest in the rock canon, a moody slice of psychedelia with nightmarish overtones. It was a tantalizing glimpse of what could be.

"Obviously the two-guitar thing with Jimmy was a great idea," Beck says. "But it was also fraught with danger, because sooner or later one of us would have been cramped, style-wise."

As it turned out, things never reached that boiling point. Two dates into a particularly grueling tour of America in October 1966, Beck walked off the bus and out of the Yardbirds, leaving Jimmy to carry the ball. Much to Page's credit, he not only endured the tour but also triumphed, marshaling the three remaining Yardbirds into a powerful, effective quartet.

In the year that followed, the Yardbirds became Page's laboratory, where he formulated much of the sound and approach that he would employ in Led Zeppelin, not to mention the swaggering onstage persona that would set a new style for rock-guitar performance. His transformation from anonymous studio musician to flamboyant rocker appeared to happen overnight: Jimmy began wearing psychedelic finery custom-made for him by Swinging London's hippest designers. He augmented his new look by retiring his sober black Les Paul in favor of a Fender Telecaster given to him by Beck, which Page hand painted in bright, kaleidoscopic colors.

The image was there, and it was very cool, but there still remained much unresolved business. Beck and Samwell-Smith had left and manager Napier-Bell also jumped ship, selling his managerial interest in the group to Peter Grant in January 1967. A burly man who stood an imposing six and

a half feet tall, Grant partnered with producer Mickie Most in an organiza-
tion called RAK Management and Production.

It was arranged that Grant and Most would take on both the Yardbirds
and Jeff Beck, who had begun his solo career. Having worked with Most in
his session days, Jimmy Page was well aware that Most was not the ideal
producer for the Yardbirds. An old-school "hit factory" guy and master of
the three-minute pop-single idiom, Most was great at making something of
manufactured teen idols like Herman's Hermits, but he was hardly the ideal
man for an evolving, experimental guitar-rock group. At the time, however,
the Yardbirds didn't have much choice.

It surprised no one that Most's ideas on how to revive the band's flagging career were very much at odds with Page's vision for the group, a vision
much more attuned to where electric guitar–driven rock was heading in the
late sixties.

The band entered De Lane Lea studios in London with Most in Febru-
ary 1967 to record the band's next single, "Little Games." In keeping with
Most's production style, the tune was penned by an outside writing team,
Phil Wainman and Harold Spiro, and was selected by the producer himself.

"Little Games," a lightweight pop tune with vaguely psychedelic overtones, became the title and lead track of the album that the Yardbirds made
shortly thereafter. The recording sessions were hurried—some accounts say
the album was made in as few as three days—and the results were a mixed
bag. The most successful tracks were those driven by Jimmy. "Smile on Me,"
the first blues number on *Little Games*, was written by Page, Relf, McCarty,
and Dreja and boasts a groove very much in the mode of guitarist Hubert
Sumlin's work with the great Howlin' Wolf. But for the song's two guitar-solo
sections, the rhythm shifts to an emphatic shuffle while Page pulls out some
of the most scorching licks he's ever committed to record.

The acoustic-guitar instrumental "White Summer" was another standout track on *Little Games*, and it served as a clear-cut harbinger of one of the
many musical elements that Led Zeppelin would bring to the table, notably
in the classic "Black Mountain Side," from the group's first album. "White
Summer" reflected Page's developing interest in Indian classical music and

alternate folk-guitar tunings, which he would mine with great success both in Zeppelin and his solo career.

But for every sublime moment like "White Summer" or the lysergic "Glimpses"—a track that featured Page's masterful latticework of watery chords, tasteful sitar, and avant-garde use of musique concrète—*Little Games* was sullied by Most's cringe-worthy pop singles, like "Ha Ha Said the Clown" and "Ten Little Indians."

"Mickie would always try to get us to record all these horrible songs," Page says. "He would say, 'Oh, c'mon, just try it! If the song is bad, we won't release it.' And of course it would always get released!"

For all of Most's efforts to recast the Yardbirds as a pop act in the studio, on the road—where Page ruled the Yardbirds' roost—Jimmy took a completely different tack, pushing the band in the experimental and hard-hitting direction he envisioned. Exemplars of this approach include "Think About It," the tough, riff-driven B-side to the band's rather tepid single "Goodnight Sweet Josephine."

Several of the milestone rock albums released in 1967 confirmed that Page's way was more in tune with what was to come. Cream's debut album, *Fresh Cream*, appeared early that year, followed by *Disraeli Gears* by year's end. Jimi Hendrix's debut disc, *Are You Experienced*, came out in May 1967. These three albums cemented the arrival of the power-trio band format and a new, heavily riff-driven mode of rock expression.

And heavy was the direction in which Jimmy Page pushed the Yardbirds as they toured across America and the world in 1967 and early 1968, logging soulful miles on countless stages even as *Little Games* tumbled precipitously and ultimately slid off the charts. Live bootleg recordings of the band from that era reveal a tight and dynamic unit capable of hitting with sledgehammer force on riff rockers like "Train Kept A-Rollin' " and conjuring unexplored musical dimensions on songs like "I'm Confused," which featured Page's newly developed technique of scraping a violin bow over his guitar strings to create a multitude of eerie and ambient textures.

Nevertheless, working against what were clearly impossible odds, Page, always as hardheaded as he was hardworking, remained determined to make the Yardbirds succeed. After all, this was the band for which he had abandoned his session career, and he stuck by them loyally to their appropriately prosaic end. Following the commercial failure of *Little Games*, the group released three more singles before finally calling it quits. The Yardbirds played their last gig at the College of Technology in the small British town of Luton on July 7, 1968.

Nevertheless, this failure hardly left Page in despair. During his time with the Yardbirds, he had formed a close alliance with Peter Grant. One afternoon, while stuck with Grant in a horrific traffic jam, Page told the manager that he had some ideas for a new band, and that this time he wanted to produce the music himself.

Grant jumped all over it.

CONVERSATION

Q:

HOW DID YOU END UP
JOINING THE YARDBIRDS?

They had asked me on a couple of other occasions to join, but Jeff Beck and I continued to casually talk about how good it could be if we were both in the group.

How I ended up actually joining the band is a pretty funny story. I went to see the Yardbirds play at this really stuffy student black-tie event at Cambridge University. The singer, Keith Relf, got quite inebriated and was being really punky. He was really staring down the establishment and put on a magnificent rock and roll performance. He was knocking things over and shouting obscenities at the audience. I really enjoyed myself, but the band's bassist, Paul Samwell-Smith, was completely incensed with Keith and his increasingly erratic behavior on the road and decided that night to leave the group.

The band had gigs coming up and they were sort of scratching their heads about who could replace Paul on such short notice. That's when I volunteered my services. I was tired of the studio grind and I figured that eventually Jeff and I would get a chance to play guitar together. Filling the mighty shoes of Paul Samwell-Smith was a bit intimidating because he was a phenomenal bass player, but I managed it.

Weren't you slightly nervous about throwing your studio gig away for what appeared to be a highly volatile group of people?
I wasn't that aware of the entire history of tension between Relf and Samwell-Smith. That wasn't something that Jeff and I would gossip about. I just thought it was about one bad gig.

The Yardbirds pioneered the idea of a rock-guitar virtuoso with Eric and Jeff. Did knowing that you'd eventually get room to shine play a part in your decision to join?
I wasn't thinking like that. As far as I was concerned, I'd already proved my point as a guitarist doing studio work and playing in the Crusaders. The idea that excited me was that Jeff and I might get a chance to really explore the possibilities of a two-guitar band. We had talked about playing harmony lines and arranging parts that would be the rock equivalent of a brass or

saxophone section from the big-band era. There wasn't really anything going like that. The closest we came to realizing our vision was on "Happenings Ten Years Time Ago."

In retrospect, the idea was really sound, but the band and managerial politics were not. I was there on wages and I was the new guy so I had very little to say about anything. Jim McCarty once said that I was so desperate to escape the session world that I would have played drums if I had to! I thought that was a bit cheeky, but he's probably right. The fact is, I was regarded as one of England's best guitarists and I left the studio to play bass!

Admittedly, it wasn't as easy as it sounds. I had to fill in for Samwell-Smith, who was considered to be one of the best bass players in the whole music scene. What is also important to understand about the Yardbirds is how great the rhythm section was. Listen to *Five Live Yardbirds* and you'll hear there's a hell of a lot going on between Paul and Jim. They were the players that really created what people called rave-ups, where the band would slowly build to the point where they sounded like they were going to explode.

The Yardbirds pioneered that dynamic build found in techno music and modern jam bands.
It was sort of trance music.

What did you think of the Eric Clapton edition of the Yardbirds?
I thought they were great, and I saw them a couple of times at the Marquee. One thing that gets overlooked when Eric was in the band was their extreme good taste when it came to choosing blues songs to cover. In those days it wasn't a good idea to do numbers that the Stones were doing. They staked out their own territory and performed the songs extremely well. For example, the band's arrangement of "I Ain't Got You" is terrific, and Eric's solo is a classic.

What did you think of the Beck edition?
With Jeff I felt they were seeking new horizons. Even when they were playing pop music, they found new things to say.

So when you came into the band you had this idea of eventually using—

Yes, two lead guitars. Jeff was going to be the primary lead guitarist, but we could see the possibilities of playing riffs in harmony and that sort of deal. If it had worked out, we definitely would've pioneered new territory with the electric guitar. It was a concept no one else was doing.

Were you disappointed when Jeff left?

I was disappointed. I was definitely disappointed. I don't know how much I want to go into this, because I don't know if I should. I championed Jeff being in the band, and that's all there is to it. The others, however, weren't having it.

So how long after you joined did you have to plan *Little Games*?

Well, it was becoming apparent that there were four Yardbirds and I was taking over the electric-guitar duties. The history has it that after Jeff left, their manager, Simon Napier-Bell, wanted to sell his interest in the brand name "Yardbirds." He had a word with Peter Grant to take over the management of Jeff Beck and the Yardbirds, and Grant decided both artists would be produced by Mickie Most, who in those days was pretty much as good as his partner. Peter and Mickie shared an office and their desks faced each other in the room, which says quite a bit as to how close they were. And that's how it went. You know, one minute we were doing a tour of America, the next minute this manager had sort of moved on his interest. I'm not saying he sold the Yardbirds, but somehow or other, that's how it was. So Jeff was going to be making solo records and we were going to do the best we could.

So going back to the original question, I already had material for *Little Games* because I had been writing and coming up with things even while I was a studio musician. "Happenings Ten Years Time Ago," for example, was basically my thing. But the biggest question was how to incorporate those ideas into the Yardbirds.

One of the central issues was Mickie Most. He was really, really good at making pop singles. His primary goal was to make hits and get in the charts. And that's sort of how it went. He had no interest whatsoever in doing albums or mixing for stereo.

So while he was focusing on the singles, we focused on making the album more of a reflection of where the group was heading. Without Jeff, I knew I had to pull out all the stops, because the Yardbirds had gained this reputation as a guitar band and I had a reputation I wanted to protect. I was keen to get some of my material, if not quite a bit of my material, in the mix so that it would get noticed. The only place to do it was either on the B-sides of singles or on the album.

On some level it was good that Mickie wasn't really interested in the album, because that gave us some freedom. On the downside, if it was an album track, you were allowed to do things in pretty much one take. You had to work fast!

We've talked about some of the less compelling tracks with Mickie Most, like the single "Ha Ha Said the Clown," but the song "Little Games" wasn't so bad.
Yeah, but it's not that convincing, either. If you were really a fan of the Yardbirds and loved what they did with the blues, "Little Games" was quite a stretch. But let's be clear, I was going along with it.

It was an uncomfortable situation for the whole band. When Mickie brought us things like "Ten Little Indians," we'd go, What the fuck is this? We were leading almost a dual life because we were going in a completely different direction live and people were really responding to it. I respected Mickie, but I started to feel we were shooting ourselves in the foot by getting anywhere near things like "Ten Little Indians." The Yardbirds was a really, really good group, and I think we made a really good go of it, but we had a lot of things working against us.

"Glimpses" is certainly one of the more interesting tracks on the album.
It was conceived as a vehicle for me to use the violin bow and taped effects onstage. The idea was sort of inspired by a demonstration stereophonic record that I had as a kid that had different sound effects like trains going by in stereo. I managed to splice a tape together of some things like that along

with the sound of the Staten Island Ferry coming out of dock. So I had this tape of all this sort of weird stuff, and it was played at the same time that we were playing "Glimpses" on the stage.

Live, it was a real sonic assault. Very heavy. That's the sort of stuff we were doing on the stage, which was quite avant-garde. It was intense stuff. I've still got the tape of all those effects!

Did you know we had [legendary Rolling Stones road manager] Ian Stewart play piano on *Little Games*? I asked him to come along. Stu was a phenomenal player, but he rarely got a chance to record. We asked him to play on "Drinking Muddy Water," and he said, "Everybody's got their own version of this," and played it gorgeously straightaway.

But to give you some sense how things were done back then, we're making an album and this is like the first track. After we finished the first take, there's this flat voice that just says, "Next." Honestly, can you imagine, we are playing this blues, having a fucking great time, we've got Stu in the studio with us, and all they have to say is . . . "Next." It was quite terrifying, really. I think I had to overdub the lead guitar on the mix, because the guitar is slightly different on the fade-out in the mono and the stereo.

To be fair to Mickie, that's how he got successful, and he was *really* successful. The Yardbirds weren't necessarily in the best of company at that point. Don't get me wrong, I'm not putting any slant on anything he did, I'm just saying it was testy for musicians that were trying to break new ground.

What was it like to work with drummer Jim McCarty? He contributed to a lot of the songwriting, which is a little unusual for a drummer.
Yeah. I liked working with Jim. I liked working with all of them, to be honest. I didn't know until I was actually in the Yardbirds that he wrote lyrics— I always thought it was Keith. But Jim did do quite a lot of lyrics. For example, on "Tinker, Tailor," I had a bit of lyrics and a chorus, but he helped fill in the verses.

Those ringing guitar parts and suspended chords on "Tinker, Tailor" are almost like a precursor to "The Song Remains the Same."

Well, it could be, except I've got two or three different demo versions, each with different guitar approaches. What is somewhat funny is I presented Mickie with the poppiest version. Here I am talking about shooting ourselves in the foot by doing pop stuff, but really I sort of enabled the situation by coming up with parts that were intentionally quite catchy. I guess I still had that instinct from doing sessions for all those years!

Let's talk a little bit about the live-show bit. I've heard different bootlegs and recordings, and what comes across is a much heavier and dynamic band from what you hear in the studio.

It was real fun to play with the Yardbirds. We evolved quite a bit from the first gig I played with them, which was in America in a Dayton, Ohio, department store! Once I got on the guitar with Jeff, I started really expressing myself. Then, after Jeff left, I stayed with the band and just kept stretching and stretching. The Yardbirds had several songs that called for lengthy improvisations, like "I'm a Man" and "Smokestack Lightning," and I took full advantage of them to develop a bunch of new ideas. After the Yardbirds fell apart and it came time to create Zeppelin, I had all those ideas as a textbook to work from. And as it was stuff I developed on my own, it was fair game for me to use some of those ideas. So both things—the studio work and the experience in the Yardbirds—were really important. They both set the scene for Zeppelin. The studio gave me discipline and an incredible working knowledge of many kinds of music, and the Yardbirds gave me time to develop my ideas.

How did you feel when the Yardbirds split?

When Keith and Jim announced they were leaving, I was disappointed because I knew that the material we were developing was really good. It wasn't like the Yardbirds material with Eric, nor was it like the Yardbirds material with Jeff, because those things were what they were.

The live gigs were really going well and the response was positive. We were becoming more esoteric and underground, but we were doing really well. And you could tell that the audiences were building—what we were playing was exactly right for what was going on, in my opinion. I just thought that we could have done a really good album. I had a lot of faith in us, but I didn't think having Mickie Most produce was necessarily very healthy for the band. I don't know what would've happened with that. But, I don't know, maybe they'd just had enough. I'm sure for them the early days of the Yardbirds were probably more attractive.

MUSICAL INTERLUDE

THE YARDBIRDS
ACCORDING TO CHRIS DREJA

HE'S ONE OF THE FEW GUITARISTS IN THE WORLD WHO
CAN SAY HE'S PLAYED IN A BAND WITH ERIC CLAPTON,
JEFF BECK, AND JIMMY PAGE. FOR FIVE YEARS, CHRIS DREJA
WAS THE QUIET BACKBONE OF THE YARDBIRDS,
AND HE SAW IT ALL GO UP . . . AND DOWN.

———————

YARDBIRDS RHYTHM GUITARIST and sometime bassist Chris Dreja happily describes himself as a "voyeur." It rings true. During his time in the band, Dreja stood in the shadows, watching and responding with his tough, scrappy chords, while Eric Clapton, Jeff Beck, and Jimmy Page soaked up the spotlight.

Dreja saw many things while in the Yardbirds. He saw his band move with astonishing rapidity from small, sweaty clubs to bigger and bigger venues. He saw his fellow musicians come, go, and occasionally self-destruct. And toward the end of his time with the band, he witnessed how Jimmy Page made the best out of a difficult situation before leaping with a stringent clarity into the future with Led Zeppelin. The following are his observations.

CHRIS DREJA Back in the early sixties, England was living in black and white. It was still crawling out from under the wreckage of World War II, but the baby boom came along and we were part of a new generation of kids that had not lived through that dreadful wartime experience. We weren't afraid of the world like our parents, and we wanted to rebuild our culture from the ground up. It was a unique period for fashion, design, music, photography, and architecture, which one cannot deny. I don't think I'm being nostalgic because history bears that out, that much of the art created from that period is still important.

While there were many factors that contributed to these changing attitudes, the British art schools in the sixties played an important role in encouraging new ways of looking at things. They were these wonderful liberal arts colleges that attracted the hip, young anti-establishment types, including Keith Relf, Jimmy Page, Eric Clapton, and people like John Lennon and Pete Townshend. You didn't have to do a lot of art, but it encouraged you to do a lot of thinking. The things we learned gave us a tremendous sense of freedom, and when you are young you don't have any fear.

When we started our bands, we did what we wanted to do—why the fuck not? Why can't you distort? Why can't you bend strings? Why can't you stick your guitar in a lavatory pail? Why can't we sound like a Gregorian chant? Why can't we play louder and faster? Truth is, we didn't think it would last, so we threw caution to the wind.

THE BRITISH MUSIC scene at the dawn of the sixties was not great. There were all these manufactured pop stars running around doing a bad job imitating the great Elvis Presley. All the songs had these well-structured middle eights and twee choruses, but they were emotionally empty. The pivotal moment came when those very few of us came across the blues from the United States. It changed everything.

When I first heard Jimmy Reed and Howlin' Wolf, I was high for weeks. I could hardly sleep. And then to have the very gall to think, Why don't we try to play this? It was pretty audacious.

Because so many of the British blues-influenced musicians came from the area around Surrey—including Page, Clapton, Beck, and the Rolling Stones—we've always jokingly referred to it as the Surrey Delta. It's particularly funny because Surrey is so genteel and English middle-class. It's about as far from Mississippi as you can get.

I was at art school with Top Topham, who was the original lead guitar player for the Yardbirds. Topham's father was in the merchant marine, and

The Yardbirds in Denmark, 1967 (© *Jorgen Angel*)

he brought back blues records from America. Eventually, we started trying to play this music that excited us. It wasn't long before we attracted like-minded musicians. Top and I started playing with drummer Jim McCarty and eventually hooked up with singer Keith Relf and bassist Paul Samwell-Smith, who were playing in a band called the Metropolitan Blues Band, or something very boring like that.

They lacked a drummer and we lacked a singer, so it made sense to get together. Soon after we amalgamated, Top left the band and we enlisted a fellow blues enthusiast named Eric Clapton. The Yardbirds instantaneously began attracting this amazingly young crowd. In a matter of weeks, we went from being a warm-up band to being the main attraction. All these great clubs sprang up out of nowhere, like the Crawdaddy and the Ricky-Tick, who obviously realized there was money to be made on this R&B thing. Even the Marquee, which had been a jazz club for many years, started booking us.

We used to do all-nighters at the Scene Club, Studio 51, Eel Pie Island—they were all happening at the time. Our big break came, however, when the

Beatles invited us to join them on a series of Christmas shows at the Hammersmith Odeon. In those days, they were doing what I would almost call a vaudeville act, with comedy skits and bits between the music.

For example, there was a long-running kids' program called *Dr. Who*, which featured these hairy alien robots called Yetis, and the Beatles would actually dress as Yetis. You could not believe this—it was really vaudeville. But of course they were huge and the opportunity allowed us to reach a bigger audience.

We had a ten- or fifteen-minute spot, and we'd play blues songs like "I Wish You Would" or "Good Morning Little School Girl," which we put out as singles. You could actually hear us because the girls screamed a little less loudly than they did for the Beatles. When the Beatles came on, it was just crazy. The girls used to throw things at them, and they didn't just throw soft things. John Lennon came up to me after one of the shows with this huge giftwrapped lump of coal that hit him. He said, "Oh, Chris, I'm not going out there again." The girls also used to throw coins—and we had *big* coins—the old English money. Boy, that was heavy. Theirs was a dangerous occupation, and you never heard them anyway because the girls were screaming like crazy during their entire act.

PEOPLE USED TO SAY, "Wow you had these amazing guitar players!" But they forget that we were all just embryos. We were only in our teens when we started. Eric used to spend days learning one little riff or learning how to bend the strings or how to stand with the guitar. All that other history was to come. By the time we did the Beatles show, he was getting that wonderful kick-in-your-face sound, and that made him our secret weapon. George Harrison was a great guitar player, but he was still playing that northern Merseybeat stuff.

Going back to the blues, there is a great story that always makes me laugh. In the UK, there was an agency called the National Jazz Federation that brought black musicians like Muddy Waters and Sonny Boy Williamson over to our country as part of a cultural exchange, and they'd hire local bands like us to back them up.

We played a bunch of shows with Sonny Boy, who must've been in his fifties at the time, and he was an evil-looking bastard. He was snaky. He was really tall and had this bulbous nose, but what a player! He used to put a chromatic harmonica in his mouth and swallow it and play. Come on now! His showmanship was something else. But he was also a drunk, and that's where I think Keith Relf unfortunately picked up his drinking habit—the briefcase with a bottle of whiskey in it. When Sonny Boy got drunk, he did the typical black-musician-playing-with-white-guys thing—he'd change things around just to mess with us. But I have to say, we were rarely caught off guard. But he did everything he could to throw us off.

Years later, Robbie Robertson of the Band told us that after Sonny Boy came back from England he told Robbie, "I played with this British band over there, and they wanted to play the blues so bad . . . and they really did play them so bad!"

AFTER THAT EXPERIENCE, we knew we weren't the real deal. We loved the blues, but we decided we needed to explore other areas of music. It was absolutely important and we wouldn't have survived otherwise. So that's when a publisher came to us with a demo tape of "For Your Love."

We had recorded a couple singles that were okay, but they hadn't made any dent in the national charts. We knew "For Your Love" was a great song, and we came up with a really progressive arrangement for it. It had bongos, upright bass, Brian Auger on Hammond organ, and everything else. I remember the studio session very clearly. It was magic and you couldn't help but notice the electricity in the air. But there was a bit of a rift going on between Paul and Eric at this point over control of the band. It was not the way that Eric wanted to go. He wanted to keep playing the blues, and he couldn't relate to "For Your Love," so he quit.

So there we are with "For Your Love" climbing up the charts like a rocket to the moon, with no lead guitar. We were in trouble because there really weren't any guitar players around in those days—guys who had the balls to bend notes and things.

The only person we could think of was Jimmy Page. We knew he was doing sessions and had a reputation for being a great, versatile player and incredibly professional, so he was an obvious choice. But it also seemed very obvious that he was comfortable doing his studio work. Who could blame him? He was learning so much. He was working with all the best people in the recording world, including producers like Glyn Johns and Mickie Most and studio musicians like Big Jim Sullivan. These were heroes.

We were in a bit of a panic, but when Jimmy turned us down we didn't blame him. He was still learning. That is why that guy is a genius producer. He was observant and he watched those other guys make and break the rules. That was a huge learning curve for him as a technician as well as a player.

But, great thing is, he turned us on to Jeff Beck, who played with a band called the Tridents. We saw them play at Eel Pie Island, and Beck was fantastic.

Jeff was really quiet. He spoke through his guitar; that's how he talked. It was probably difficult for him because he was joining a band that had already been together for a few years. And, you know, bands are weird things. They're closer than marriages. We had our own language, our own humor—a lot of it invented by Eric, if I may say so. Jeff was different from the rest of us and did look a bit rough—he was a mechanic. I remember him telling me this story about how someone pissed him off and he poured paint stripper all over their car. I thought, [sarcastically] "Yes, this is just what we need in this band." Jeff wasn't a great fit socially, but he was always playing the most amazing music.

So Jeff comes into this readymade outfit with "For Your Love" climbing the charts. Suddenly we're on theater tours, we're on television, and because I had learned how to shop for clothes from Eric, who was a very dapper dresser, I was given the task of cleaning Jeff up. I took him off to Carnaby Street and he got a pucker haircut and we bought him nice shirts and things.

We always knew that Jeff was a genius, almost from the word go. He wasn't a traditionalist like Eric; he was a great experimenter. If you wanted a sound like a police siren, Jeff would make it. He'd make it happen. Clucking chickens. Sitars. Whatever!

On *Roger the Engineer,* we'd write the music, put the backing track down, and then bring Jeff in to crawl all over it. And he always came up with these great sounds. We never thought he couldn't. I remember reading an interview where Jeff said the Yardbirds used to put a lot of pressure on him because they wanted all this stuff and he had to come up with it. Of course we never knew that he might have had that angst at all. There was that famous session for "Heart Full of Soul" where we got tabla players and a sitar player to play the main riff, but it was hopeless because they couldn't get the timing right. Jeff came with a fuzz box and said, "Well, why don't we do it *this* way?" And, of course, he nailed it.

He was a genius at creating soundscapes. I don't think we appreciated how good he was, because he was brilliant.

OUR SINGER, Keith Relf, was a great original. I think his songwriting was very strong. And just listen to his harmonica playing—no one was playing harmonica like that, no one. No one was riffing with a guitar player like he did. I think Keith's vocal interpretations of "Heart Full of Soul" and "For Your Love" were brilliant as well. Okay, you can get stronger singers like Robert Plant, but he was a real craftsman and an original.

Unfortunately, he was a heavy drinker and self-destructive. I knew he was going to die young. He had terrible asthma and actually lost a lung during his tenure with us while Eric was still in the band. Can you imagine: a singer that played harmonica, with only one lung? To add to his misery, we didn't have good PAs back then. Vox supplied us with PA columns, but quite frankly you could hear fuck all. It was a nightmare for a singer and a nightmare for a singer who also played harmonica.

Guitar players had their own amplifiers, and those early Vox amps made a huge sound. Jeff had an ego and he liked it loud, and he needed it loud so he could do his effects. Now, that's going to kill a singer back then to compete with that. I know it was difficult for Keith. He did get kind of flattened. In the process of trying to keep up, Keith often lost his voice.

He was aware of his image, but he was not a sexual human being in that sense that he padded his codpiece. It didn't take too long before the power of a guitar player like Beck overshadowed him.

PAUL SAMWELL-SMITH was almost the complete opposite of Keith and Jeff—he was sort of straitlaced and very un-rock and roll. I knew he was getting very tired of the road. He found it very uncivilized, and it *was* uncivilized in those days. The people who ran the promotions, especially in America, were mainly Mafia. I remember playing Vanilla Fudge's club in Long Island and being introduced to truly menacing people who were eight feet tall, with chewed-up ears and smashed-in faces, and had names like Vinnie.

The Yardbirds always found ourselves in the most amazing circumstances. We played some rough gigs, man. Once in Wales we played this sort of huge lavatory, and the promoter said, "You start at ten, but whatever happens, *don't stop playing* until we tell you to stop." And we're thinking, What the fuck is going to happen? There was nobody in the place—nobody! Then at eleven o'clock, the girls started coming in. They were pissed out of their brains, stumbling around and vomiting on the walls. And then at eleven thirty, the pubs closed and the guys came in and they just started fighting each other. One guy threw a chair and smashed this guy, who was standing in front of Eric, in the face. Eric had just bought a new white Telecaster, and this huge stream of blood came flying across the room and splattered all over the front of it. We were playing a Chuck Berry number, and we're both saying to each other, "Whatever you do, keep playing . . . *for God's sake, keep playing!*"

The police eventually came and broke it up. Apparently this is what happened every Friday night in this town. Two rival coal-mining communities got together and they smashed the shit out of each other. Of course, not every gig was like that, but the Yardbirds played every type of gig you could imagine, from universities to stadiums to theaters to cinemas to intimate little clubs that could only fit a hundred people.

Paul wasn't happy with any of this, which brings us to how he left the band. Paul was a little bit of a snob, and he liked the established classes a little bit. He was pleased when we were hired to play the Cambridge University

May Ball, which was very prestigious and upper-class. They only booked big artists and they had lots of money. It was a bit of a stuffy event, but they had amazing catering for the artists that included all the wine and food we could consume. Keith was heavily into his drinking stage by this time, and I think he felt a little uncomfortable playing to the elite of the English establishment, who were awfully stiff. I mean, these people didn't really know how to even dance—they were like robots, it was so funny.

Keith got seriously drunk. He got so drunk that backstage he and Graham Nash from the Hollies, who were also on the bill, were karate-chopping plastic trays. They started off with one and ended up with five. Of course, what did he do? He broke all his fingers on his right hand. He was so pissed he didn't even realize what had happened. During our last set we had to literally tie him to the microphone. And all he could sing was [*makes farting sounds*]. Man, this was punk. This was real, genuine punk. And unbelievably or not, on this day, Jeff had brought Jimmy along. Jimmy loved it. He just thought it was the best show he'd ever seen.

As much as Jimmy loved it, Paul hated it. So there we are again: Paul leaves, and once again we're looking for a new band member. So we turned to Jimmy again, and this time the timing was perfect. In a way, it wasn't surprising he joined. The Yardbirds were a band that a guitar player wanted to be involved in. There was no other British band, really, that would let you spread your wings. Jimmy was tired of being a studio musician and leapt at the chance, so much so that he came in on bass, briefly. Right away, he loved being back in with a live band. He's a clever guy, and he saw the future moving from the studio to the concert stage, and by this time the Yardbirds could play pretty much anywhere in the world.

So, yeah, we got Jimmy, and on paper we had a dream band with two top lead guitarists. We could've had the British version of the Allman Brothers . . . before the Allman Brothers. But we didn't have that. Jimmy was highly professional, but he also had an ego. And we had a huge ego in Jeff, who we didn't realize by that time had very much bathed in his own perfect spotlight. Who was going to take it away from Jeff? Well, there was only one bloke who was going to take it away from Jeff, and that was Jimmy.

But as soon as they started working together on guitar, it clicked. You can hear how it would've worked on one of my favorite singles, "Happenings Ten Years Time Ago." It's a perfect two-and-a-half-minute rock opera.

Ultimately, I think Jeff felt a little deflated that he had to share the spotlight. He was fine sharing the stage with me because I was no threat. I was there to make Jeff sound good, and with me playing rhythm you always will sound good. But when Jimmy came along with his talent and his energy, Jeff felt threatened. Jimmy did bring that energy in.

I think Jeff left for a combination of reasons. He did feel a bit intimidated by having this other great guitar player on the scene, and he was sick of the touring conditions. Our manager, Simon Napier-Bell, had put us on this dreadful Dick Clark tour that used these old Greyhound buses to transport us around, and that's going to kill anybody off. Halfway through that tour it was plicky-plunky. And Jeff threw a wobbly somewhere in the middle of it, smashed a guitar in front of me in the dressing room and went back to L.A. Suddenly we were a four-piece. Being a professional, Jimmy said, "We've got a contract, let's carry on playing." He started doing much more on the guitar and seized the moment.

It was one of the best times in the band for me. The early years were exciting, but it got difficult by about the seventh American tour. Then when Jimmy came along and I shifted to bass, it became fun again. Jim and Keith, unfortunately, were going the other way. They were dabbling in drugs, and occasionally they'd disappear.

When the band became a quartet, we split into two camps. There was Jim McCarty and Keith, who used to travel together in their Mini Cooper, and Jimmy and I used to travel in my Mini. Jimmy couldn't drive, but I loved driving, so it didn't matter. I owned a Mini Cooper S, which was so light and so ridiculously overpowered. Stupid car—but what a drive! We'd come back from gigs late at night, and back then there were hardly any motorways. You'd have to drive miles through country lanes. Jimmy doesn't know that I nearly killed him. I never told him this story, but I was coming around this corner out of a village at eighty or ninety miles an hour, and there was a fucking donkey in the road. He's asleep. I missed this donkey by this much.

We would have been legends in our own death. And of course, I always laugh because he used to go to sleep after the gig, and I used to swerve to see how much I could make his head bang back and forth before I'd wake him up!

Unfortunately, *Little Games*, the album we made with Jimmy, was a throwaway project for our producer, Mickie Most, and it didn't reflect where we were going or our live show at all. Mickie was a hit maker, and he really didn't get us. I remember in his office, he'd pull out a drawer and there were all these demo tapes in it, and of course nothing he brought out was really good for us, apart from maybe "Little Games."

Mickie was very successful, but he wanted the backing track at ten, lunch at twelve, overdubs vocal by five, and home for dinner. All he cared about were singles and he didn't give a shit about albums. Basically, he said, "You guys can do whatever you want on the album, as long as you do it in a certain amount of time." Well, once again, we had a secret weapon: We had Jimmy's technical skill. So that album was written and recorded pretty much by us and for my money, it's the best thing that ever came out of the Mickie Most era. It's rough at times, and it's big at times, but it's ahead of its time. It's a cult album. Like "Happenings Ten Years Time Ago," it took a while for it to be brought to the heart.

You can see how clever Jimmy was in those days. Even when he was in the Yardbirds, he knew that times were changing and you didn't need a single. He knew that albums were becoming more important and he was more focused on trying to capture the full flavor of the band. He was able to realize that vision with Led Zeppelin, and he proved that he was right.

I'm not one for regret, but I wished the Yardbirds would have continued. I would have walked through those doors with Jimmy as the Yardbirds. I thought we were a great four-piece. We were getting bigger, fatter, and creative in a different way. Jimmy brought that life into the band. But by that point we had two guys that had had enough. They wanted to play what I would call "water music," or New Age music. It did nothing for me, but that's what they wanted to do. So they signed off on our last tour through an attorney, saying no more tours and thank you very much.

Jimmy was disappointed, but he walked away with eighteen months of ideas in his pocket. I was a bit road weary. I joined the band at fifteen and left at twenty-one, so I was still very young. I only ever wanted to play with the Yardbirds, and I had this other great love— photography. I didn't know what the future would hold, but I was tired from waking up every morning and relying on an alcoholic or a druggie. I couldn't do that anymore. I wanted to shape my own destiny for a while, and photography suited that bill entirely.

I knew that Jimmy wanted to carry on, and the Yardbirds had some dates in Scandinavia that needed to be fulfilled. Obviously, the Yardbirds didn't exist, but Jimmy wanted to put a band together and do the dates. I went up to Birmingham with Peter and Jimmy to have a look at Robert Plant. He was playing with the great John Bonham, and we all said, That's the drummer you need, Jimmy. Funnily enough, no one was quite sure about Robert Plant, because he was a little bit of a shrieker in those days.

John Paul Jones was a great bass player, and he had such a terrific sound. He was using those early Ampegs and the Fender Jazz bass, whereas we all played the Rivoli, which has a bit of a wallowy sound. He had this great, clicky, perfect sound, and there couldn't have been a better bass player for Led Zeppelin. So at the end of the day, Jimmy ended up putting together the perfect combination, and he'd worked up loads of ideas from the Yardbirds, so he had it all up in his brain. They sat around on their first audition playing "Train Kept A-Rollin' " or whatever, and if you click as musicians playing that, then you know you've got something. So I was out of it. Very relieved at the time. I had no money. I was twenty-one, and I think I had about three hundred dollars in the bank. It's crazy, isn't it? But it didn't matter. I had photography and I made it work.

I went to work in New York soon after, and I think I was only recognized once and it was by a mail messenger. My studio was on Fifth Avenue, near Washington Square, and this guy came in with his envelopes. He was in the studio while we were shooting, and he said, "Aren't you Chris Dreja from the Yardbirds?" Nobody in my world knew that I'd been a musician and I never mentioned it at all.

I was so traumatized after the split, I couldn't even listen to music for a few years. I went to New York, and in the meantime, Jimmy had put Led

Zeppelin together and it became a monster. Jimmy let me shoot their rear album cover for their first album. They paid me twenty-one guineas. Jimmy and I had this great relationship about photographs because he trusted me implicitly. But also I used to make him look good!

One day, while my wife and I were living in Brooklyn, Peter Grant rang me up to say the band was playing Madison Square Garden and wondered if I wanted to stop by. I thought it'd be nice to see everybody again, so we met downstairs in the concrete car park and they took me to the dressing room. They were nice, nice guys to me—no ego, and very reverent in many ways. I remember Jimmy turning to me and saying, "We're going to go out and play now, Chris, you join us. Sit where you like." I walked out, and at this time they used to kick off their set with "Whole Lotta Love." And I remember walking up the concrete to the backstage area, and the building was moving—all that concrete was moving. I walked up to the biggest fucking rock and roll sound I'd ever heard, and the whole thing was so huge, I couldn't believe it. I was so out of it I didn't realize how big the gig was going to be.

When I left the Yardbirds, our biggest audience was maybe five thousand people. Suddenly, here were twenty-five to thirty thousand people. I'd been like a monk in that interim period. What a revelation.

[CHAPTER]

PAGE LEAVES THE YARDBIRDS,
FORMS LED ZEPPELIN, AND RECORDS
THEIR FIRST TWO ALBUMS.

Page performing overdubs on "Whole Lotta Love" at A&M Studios, Los Angeles, 1969
(© Chuck Boyd)

"I WANTED ARTISTIC CONTROL
IN A VISE GRIP . . ."

EARLY IN THE summer of 1968, Jimmy Page retreated to his renovated Victorian boathouse on the Thames River to plot his next move. The Yardbirds had broken up, but so be it. It was the late sixties—one of the most exciting periods in pop-music history—and culture was changing at the speed of sound. It was a time for action, and Page knew exactly what he wanted to do.

The Yardbirds had toured America extensively, and that experience allowed the guitarist to get in tune with the evolving tastes of the U.S. market. "In the late sixties, American FM radio was very free-form and would play entire sides of albums, including the more experimental bands like the Yardbirds, Cream, and Traffic. I knew I could create something special for that musical landscape," he recalls.

To that end, the guitarist carefully devised a blueprint for his ideal band. The group that he envisioned would "combine blues, hard rock, and acoustic music, topped off with memorably heavy choruses."

In a stroke of cosmic luck, he rather quickly found just the right men to help him achieve his ambitious musical goals. John Paul Jones, one of England's finest session bassists, keyboardists, and arrangers, was just as tired of the studio grind as Jimmy once was and asked to join Page's new band. Recognizing Jones's vast talent and versatility, Jimmy immediately enlisted the multi-instrumentalist.

Vocalist Robert Plant, recommended by Page's friend, singer Terry Reid, was also a remarkable find. The leonine Plant not only looked every inch the lead singer, but his volcanic, androgynous voice was suitable for everything from the raunchiest blues to the most delicate ballad.

But perhaps his greatest discovery was drummer John Bonham, who had played with Plant in previous bands. Page had wanted a very powerful drummer, but Bonham was "beyond the realms of anything I could have possibly imagined," he says. "He was superhuman."

From their first rehearsal in a small basement room on Gerrard Street in what is now Chinatown in London, it was evident that the band was going to work. The room exploded and Page remembers the four musicians just laughing at how good they sounded jamming on songs like "Train Kept A-Rollin'," which had been made famous by the Yardbirds. There are contradicting tales regarding how the band settled on the name Led Zeppelin. The most persistent is that Who drummer Keith Moon came up with the phrase while Page was producing the "Beck's Bolero" session featuring himself, Moon, Jeff Beck, John Paul Jones, and pianist Nicky Hopkins several months earlier (see page 27).

"Keith was all fired up at the session and said we should form a permanent band and call it Led Zeppelin," Page says. "It was a twist on the expression about a bad joke going over like a lead balloon. It stuck with me because I thought it was funny, and I liked the 'heavy and light' connotation."

With his dream band assembled, Page was so confident that he decided to produce the group himself and sell the results to the highest-bidding record label. Led Zeppelin assembled in London's Olympic Studios in November 1968 with Page producing, and after only a few weeks of rehearsal and a short tour of Scandinavia, they cut their groundbreaking debut in a mere thirty hours at a cost of roughly £1,700.

But even on this tight schedule, the band, just as Page had envisioned, produced plenty of "light and shade." From the ominous "Dazed and

Confused" to the folky "Babe I'm Gonna Leave You" to the quicksilver pop of "Communication Breakdown," Zep's power, versatility, and imagination were undeniable.

Not surprisingly, at front and center of *Led Zeppelin* was Page's innovative guitar playing. All the years of performing in clubs, playing countless studio dates, jamming with Britain's best musicians, and developing his own unique voice in the Yardbirds paid off in a diverse and sophisticated showcase of power and subtlety. In many ways, Led Zeppelin's debut made a convincing case that Jimmy was rock's consummate player. He could be psychedelic ("How Many More Times"), bluesy ("You Shook Me"), play acoustic in the style of the day's leading folk revivalists ("Black Mountain Side"), and, perhaps most important of all, be a true revolutionary.

On one segment of the album's masterwork, "Dazed and Confused," Page famously played his Telecaster with a violin bow, creating a sinister effect that even today causes the listener to sit up and take notice. Developed while Jimmy was still in the Yardbirds, the dramatic composition is structured so that each member of the band is given ample opportunity to demonstrate his unique abilities. John Paul Jones's walking bass rumbles hauntingly, Bonham's drums explode like a series of mini-volcanoes, and Robert Plant howls and moans like a man who is simultaneously in the throes of exquisitely agonizing torment and hard-core sexual ecstasy.

Page's guitar, however, is the true wonder here. Recorded primarily in a single take using just his Telecaster, a Vox amp, a Sola Sound Tone Bender for distortion, a wah-wah pedal, and a violin bow, the guitar sounds like an orchestra of otherworldly textures and sonic dread. So titanic was the composition that it would become Jimmy's signature and the cornerstone for now-legendary group improvisations that would go on for well over twenty minutes at many gigs.

Led Zeppelin's debut also demonstrated that Page was a producer to be reckoned with. Filled with incredible performances, grand compositions, and dazzling effects such as the backward echo on "You Shook Me," the record made it clear that Jimmy knew how to get the best out of his band and could do so in a disciplined, speedy fashion.

With the album finished, Page enlisted former Yardbirds manager Peter Grant to take Led Zeppelin to the bank. Using the guitarist's association with the Yardbirds as bait, the ruthless Grant worked out a five-year world-wide distribution deal with Atlantic Records. Under the landmark terms, the band was promised total creative control—their records would be produced independently, without any label interference. The group would also control all jacket artwork, press ads, publicity pictures, and anything else related to their image.

On January 12, 1969, *Led Zeppelin* was released. It entered the American album charts at number 99. Its highest position was number 10, but the album remained on the charts for a remarkable seventy-three consecutive weeks. While Led Zeppelin's impact was unusual for a new band, especially impressive was the fact that the group's popularity was the product of word-of-mouth enthusiasm, underground radio exposure, and an unstoppable live show. No singles from their debut were released to top-20 radio stations, as was almost always the case, and the band had practically no support from the rock press, which was caught off guard by the band's overnight success.

As it turned out, Page's prediction regarding the powerful rise of American FM radio had been right on the money.

AFTER THE RUNAWAY success of the first album, Led Zeppelin was faced with an interesting dilemma. Should the band stay on the road to promote sales of its debut, or return to the studio and quickly knock out a second album while the iron was still hot? For Jimmy Page and manager Peter Grant, the answer was obvious and outrageous: They opted to do both.

While the decision to record while on tour was somewhat unorthodox, it was not quite as reckless as it sounded. The last thing Zeppelin wanted to do was leave the road, for it was widely acknowledged that it was the band's incredibly dynamic live show that was kicking the first album into orbit.

At the time the band broke, the hippie movement was in full bloom and the concert circuit was crowded with shaggy, self-indulgent jam bands with little or no stage presence. In stark contrast, Zeppelin set out to hit people hard and put on a show. They wanted to look good, sound good, and play better. And at a time when few artists placed a premium on entertainment—to play was the thing—Page and his cohorts electrified audiences with their dynamic, extremely focused act.

"When Led Zeppelin first came out I thought they were fantastic," Who singer Roger Daltrey told *Classic Rock* magazine. "They supported us on one their first gigs in the States in Maryland. I stood on the side of the stage and watched their set, and I thought they were brilliant. I was impressed with the whole band. We obviously knew the guys, and I knew Jimmy from way back. He played [as a session musician] on the Who's first single. They were like Cream, but with a lot more weight. Jack Bruce of the Cream was really a jazz and blues singer, but Robert knew how to rock.

"Throughout our early history, we used to do loads of gigs with Hendrix and Cream, that three-piece-band-and-a-singer formula. We were well schooled in that, but Zeppelin took it to another level. There was a power there. All of a sudden, this was a new form of music. The music scene was starting to get a bit tired. Even Hendrix was starting to get tired then. He was moving into jazz. Zeppelin regenerated it."

Writing and recording while on the road would be difficult, but Page and the band were determined to make it work. With their ambition and creativity firing on all cylinders, the quartet would make its second album bigger and better than its predecessor. "Too many groups sit back after their first album, and the second one is a downer," Page remarked to Zeppelin biographer Ritchie Yorke. "I wanted to make every album better than the last—that's the whole point of it."

The band went more than one step further. Page and company hopped, skipped, and jumped all over the world to meet their goal of finishing the album even as they braved a brutally hectic tour schedule. But while *Led Zeppelin II* was recorded and mixed in various studios in London, New York, Memphis, Los Angeles, and Vancouver throughout 1969, it sounded anything but disjointed. In fact, the album was exactly what Page wanted—a consolidation of the best ideas found on Zeppelin's first album, with just enough experimentation to show that the band was determined to grow, as in the jazzy "What Is and What Should Never Be" or the unprecedented sexual tsunami that was "Whole Lotta Love."

Like the first album, Zeppelin's new work crackled with the excitement and energy of discovery. There were, however, some significant changes. While on tour in early 1969, Jimmy retired his 1958 Telecaster and started playing a 1959 sunburst Les Paul Standard that he bought from James Gang and future Eagles guitarist Joe Walsh.

In the late sixties, Walsh attended Kent State University in Ohio while playing in various bands in the Cleveland area. He had met Page while Jimmy was still in the Yardbirds, and they met up again when the James Gang opened for Zeppelin in early 1969.

"I always thought Jimmy was a great guy and a great player," said Walsh. "When I first met him he was still playing the dragon Telecaster, but he told me he was looking for a new guitar with more balls. I happenend to own two Les Pauls at the time and I offered him the one that I liked a little less! The guitar was quite distinctive because the neck was too fat for my taste, so I had it sanded down and reshaped by the late Virgil Lay, the legendary luthier and owner of Lay's Guitar Shop in Akron, Ohio.

"I wasn't really a guitar collector the way people are now, but I used to love digging through pawnshops and small music stores for cool guitars and amps that I would fix up. Les Pauls weren't really cost prohibitive back then, but they were hard to find. Eric Clapton played one on *Blues Breakers with Eric Clapton* which was really influential. After that everyone wanted a Les Paul and they all just started disappearing."

"It was a guitar I was meant to have," Page says. "Joe Walsh told me I should buy this guitar. He was right. It became my wife and mistress . . . without the alimony!" Page also changed amplifiers, settling on the new 100-watt amplifiers made by Marshall. He explained in 1998, "It was state-of-the-art reliability. They sounded great and were dependable on the road. I was always having trouble with amps—fuses blowing and whatnot."

The combination of the new guitar and amp created the bigger, denser sound heard on the thunderous, riff-driven "Heartbreaker," which featured a stunningly fast and complex unaccompanied guitar cadenza that eventually became a Page concert highlight.

"Lemon Song," recorded in May at Mystic Studios in Hollywood during the group's second tour of the United States, was also a standout. While the recording sounds gargantuan, in reality it was captured completely live in a tiny sixteen-by-sixteen-foot studio. "The room, where Ritchie Valens and Bobby Fuller once recorded, had wooden walls and lots of ambience," Page says. "It was a small room, but the energy of the band really comes through."

"I saw Zeppelin in the summer between the release of their first and second album," recalls Kiss guitarist Paul Stanley. "They were the most astonishing band I've seen to this day. My friend and I left the concert and looked at each other and said, 'Let's not say anything, it'll only cheapen it.' It was the perfect marriage of all the elements that made great rock and roll. It was sexy, ruthless, and dangerous."

Like its predecessor, *Led Zeppelin II* was an instant smash. Eventually it would knock the Beatles' *Abbey Road* off the top of the U.S. charts, where it would reside for seven weeks before being dislodged by Simon and Garfunkel's classic *Bridge Over Troubled Water.*

It is often suggested that the first two Zeppelin albums marked the crystallization of heavy metal. The truth is, they were far more significant. It could be argued that *Zeppelin I* and *II* ushered in a whole new era of music, with the impact of the two albums growing stronger with each passing year, particularly in the production of hip-hop and modern rock and pop.

Until Zeppelin, drums on most albums were relegated to a supporting role. On many records made in the sixties, the kick drum—if not the whole kit—was merely audible. A number of satisfactory explanations exist for this: Some of it was due to the limitations of recording technology, some was due to the incompetence of recording engineers, and a great deal was due to the simple fact that it was the status quo. Page, who was seeking to maximize Zeppelin's impact, began looking for an entirely new way to record drums.

The results of his experiments were so impressive that Page now claims that one of his greatest achievements as a producer/engineer was the way he miked drummer John "Bonzo" Bonham's kit.

The power of the drums on *Led Zeppelin II* is indeed astonishing. The snare drum virtually explodes on "Whole Lotta Love," while the bass drum kicks like a team of angry horses on "Heartbreaker" and every other track. Bonham throws the notion of a drummer as a mere timekeeper out the window on *II*. On every track, his sound is as important and present as the vocals or the guitars. As if to emphasize the point, Page makes Bonham's drumming and drum sound the clear focus of the instrumental track "Moby Dick."

After *Zeppelin II*, the band closed the door to any thought of going back to pop music. Page and Bonham had opened a Pandora's box of percussive possibilities, and since the release of the first two Zep albums, the role of drums and the sonic space they occupy in music has continued to evolve. From the larger-than-life tom-tom sound that drove the bombastic pop power ballads of the eighties to the Beastie Boys' kick-'n'-snare-propelled hip-hop to Metallica's thundering thrash, almost every contemporary recording artist owes some debt of gratitude to Page's pioneering drumscapes.

CONVERSATION

Q:

RIGHT FROM THE BEGINNING, YOU WERE
ABLE TO TRANSLATE THE EXTREME
DYNAMISM OF LED ZEPPELIN'S LIVE ACT
INTO A DYNAMIC STUDIO RECORDING.
WHAT WAS YOUR SECRET?

That *is* interesting, isn't it? One usually thinks of a dynamic album being translated into a dynamic live performance, but in the early days, it was the other way around for us. I think part of the key was that we miked John Bonham's drums like a proper acoustic instrument in a good acoustic environment. The drums had to sound good because they were going to be the backbone of the band. So I worked hard on microphone placement. But then again, you see, when you have someone who is as powerful as John Bonham going for you, the battle is all but won.

So the way to capture a dynamic performance is, essentially, to capture the natural sound of the instruments.
Sure. You shouldn't really have to use EQ in the studio if the instruments sound good. It should all be done with microphones and microphone placement. The instruments that bleed into each other are what creates the ambience. Once you start cleaning everything up, you lose it. You lose that sort of halo that bleeding creates. Then if you eliminate the halo, you have to go back and put in some artificial reverb, which is never as good.

That's particularly true of the blues. Playing the blues is not a cerebral experience. It's often said that musicians try to summon the spirit of the blues when they play. But how can that be done when there is no room, no space? And space is also essential if you want to capture the mojo created by musicians playing together live.
And that's probably the biggest difference between the music made in the fifties and music made from the seventies on—everything suddenly had to be cleaned up. You do that and you take that whole punch out of the track.

Along with that strong musical vision you had in those early days, you also took a unique approach to handling the business aspect of the band. By producing the first album and tour yourself, was it your intention to keep record-company interference to a minimum and maximize the band's artistic control?
That's true. I wanted artistic control in a vise grip, because I knew exactly what I wanted to do with the band. In fact, I financed and completely

recorded the first album before going to Atlantic. It wasn't your typical story where you get an advance to make an album—we arrived at Atlantic with tapes in hand. The other advantage to having such a clear vision of what I wanted the band to be was that it kept recording costs to a minimum. We recorded the whole first album in a matter of thirty hours. That's the truth. I know, because I paid the bill. [*laughs*] But it wasn't all that difficult because we were well rehearsed, having just finished a tour of Scandinavia, and I knew exactly what I wanted to do in every respect. I knew where all the guitars were going to go and how it was going to sound—everything.

The stereo mixes on the first two albums were very innovative. Was this planned before you entered the studio as well?

I wouldn't go that far. Certainly, though, after the overdubs were completed I had an idea of the stereo picture and where the echo returns would be. For example, on "How Many More Times," you'll notice there are instances where the guitar is on one side and the echo return is on the other. Those things were my ideas. I would say the only real problem we had with the first album was leakage from the vocals. Robert's voice was extremely powerful and, as a result, would get on some of the other tracks. But oddly, the leakage sounds intentional. I was very good at salvaging things that went wrong.

For example, the rhythm track in the beginning of "Celebration Day" [from *Led Zeppelin III*] was completely wiped by an engineer. I forget what we were recording, but I was listening through the headphones and nothing was coming through. I started yelling, "What the hell is going on?" Then I noticed that the red recording light was on what used to be the drums. The engineer had accidentally recorded over Bonzo! And that is why you have that synthesizer drone from the end of "Friends" going into "Celebration Day," until the rhythm track catches up. We put that on to compensate for the missing drum track. That's called salvaging.

The idea of having a grand vision and sticking to it is more characteristic of the fine arts than of rock music. Did your having attended art school influence your thinking?

No doubt about it. One thing I discovered was that most of the abstract paint-ers that I admired were also very good technical draftsman. Each had spent long periods of time being an apprentice and learning the fundamentals of classical composition and painting before they went off to do their own thing.

This made an impact on me because I could see I was running on a par-allel path with my music. Playing in my early bands, working as a studio musician, producing, and going to art school was, in retrospect, my appren-ticeship. I was learning and creating a solid foundation of ideas, but I wasn't *really* playing music. Then I joined the Yardbirds, and suddenly—*bang!*—all that I had learned began to fall into place, and I was off and ready to do some-thing interesting. I had a voracious appetite for this new feeling of confidence.

You've often described your music in terms of "light and shade," which are defi-nitions used in painting and photography rather than rock music.
"Structure," as well; architecture plays a part, too.

"Good Times Bad Times" kicks off Led Zeppelin. What do you remember about recording that particular track?
The most stunning thing about that track, of course, is Bonzo's amazing kick drum. It's superhuman when you realize he was not playing with a double kick. That's one kick drum! That's when people started understand-ing what he was all about.

What did you use to overdrive the Leslie [a rotating speaker used primarily for organs] on the solo?
[*thinks hard*] You know . . . I don't remember what I used on "Good Times Bad Times." But curiously, I do remember using the board to overdrive a Leslie cabinet for the main riff in "How Many More Times." It doesn't sound like a Leslie because I wasn't employing the rotating speakers. Surprisingly, that sound has real weight. The guitar is going through the board, then through an amp that was driving the Leslie cabinet. It was a very successful experiment.

How did you develop the backward echo at the end of "You Shook Me"?

When I was still in the Yardbirds, our producer, Mickie Most, would always try to get us to record all these horrible songs. During one session we recorded "Ten Little Indians," an extremely silly song that featured a truly awful brass arrangement. In fact, the whole track sounded terrible. In a desperate attempt to salvage it, I hit upon an idea. I said, "Look, turn the tape over and employ the echo for the brass on a spare track. Then turn it back over and we'll get the echo preceding the signal." The result was very interesting—it made the track sound like it was going backward.

Later, when we recorded "You Shook Me," I told the engineer, Glyn Johns, that I wanted to use backward echo on the end. He said, "Jimmy, it can't be done." I said, "Yes, it can. I've already done it." Then he began arguing, so I said, "Look, I'm the producer. I'm going to tell you what to do, and just do it." So he grudgingly did everything I told him to, and when we were finished he started refusing to push the fader up so we could hear the result. Finally I had to scream, "Push the bloody fader up!" And lo and behold, the effect worked perfectly. When Glyn heard the result, he looked bloody ill. He just couldn't accept that someone knew something that he didn't—especially a musician.

The funny thing is Glyn did the next Stones album, and what was on it? Backward echo! And I'm sure he took full credit for the effect.

When people talk about early Zeppelin, they tend to focus on the band's heavier aspects. But your secret weapon was your ability to write great hooks. "Good Times Bad Times" has a classic pop hook. Did playing sessions in your pre-Yardbirds days hone your ability to write memorable parts?

I would say so. I learned things even on my worst sessions—and believe me, I played on some horrendous things.

Did your friends ever tease you for playing jingles?

I never told them what I was doing. I've got a lot of skeletons in my closet, I'll tell ya!

How did "How Many More Times" evolve?

That has the kitchen sink on it, doesn't it? It was made up of little pieces I developed when I was with the Yardbirds, as were other numbers such as "Dazed and Confused." It was played live in the studio with cues and nods.

John Bonham received songwriting credit for "How Many More Times." What was his role?

I initiated most of the changes and riffs, but if something was derived from the blues, I tried to split the credit between band members. [Robert Plant did not receive any songwriting credits on *Led Zeppelin*, as he was still under contract to CBS.] And that was fair, especially if any of the fellows had input on the arrangement.

You also used a violin bow on the strings of your guitar on that track.

Yes, like I said, we used the kitchen sink. I think I did some good things with the bow on that track, but I really got much better with it later on. For example, I think there is some really serious bow playing on the live album [*The Song Remains the Same*]. I think some of the melodic lines are pretty incredible. I remember being really surprised with it when I heard it played back. I thought, Boy, that really was an innovation that meant something.

How did you come up with the idea of using a violin bow on an electric guitar?

When I was a session musician, I would often play with string sections. For the most part, the string players would keep to themselves, except for a guy who one day asked me if I ever thought of playing my guitar with a bow. I said I didn't think it would work because the bridge of the guitar isn't arched like it is on a violin or cello. But he insisted that I give it a try, and he gave me his bow. And whatever squeaks I made sort of intrigued me. I didn't really start developing the technique for quite some time later, but he was the guy that turned me on to the idea.

Your bow playing, especially on "Dazed and Confused," is really enhanced by echo.

It was actually reverb. We used those old EMT plate reverbs.

That's a little surprising, because there are areas that sound like you're using tape echo. In fact, *Led Zeppelin* was the first album that I can think of that employed such long echoes and delays.

It's a little difficult to remember, and I can't tell you on exactly which tracks, but there was a lot of EMT plate reverb put onto tape and then delayed—machine delayed. You were only given so much time on those old spring reverbs.

Another interesting aspect of the first album is your use of the acoustic guitar, which was something that separated you from other guitar heroes of the day like Clapton or Hendrix.

Our acoustic songs were designed to create dynamics both on the albums and in live performance. The harder songs wouldn't have had as much impact without the softer ones. It was funny to us that everyone made such a big deal out of the fact that we used acoustic instruments on *Led Zeppelin III*, because they were there from the beginning. The first album featured two folk-oriented songs, "Babe I'm Gonna Leave You" and "Black Mountain Side," but the critics just never noticed. It makes you wonder. I think they just got completely absorbed in the second album, which was more high-energy. But even the second album had its quiet moments in "Ramble On" and "Thank You."

Our performance of "Babe I'm Gonna Leave You" showed how original the band was. There weren't many hard-rock groups who would have the nerve to play a song originally recorded by Joan Baez!

Your arrangement of the traditional British folk song "Black Mountain Side" was also an interesting choice. It was a tune recorded by revivalists like Anne Briggs and Bert Jansch, but you saw the potential as a rock song.

As a musician, I'm only the product of my influences. The fact that I was listening to folk, classical, and Indian music in addition to rock and blues was one thing that set me apart from so many other guitarists at the time.

How did the Indian and Middle Eastern elements heard on songs like "Black Mountain Side," "White Summer," "Friends," and "Kashmir" take form?
Eastern music has always appealed to me. I went to India after I came back from a tour with the Yardbirds in the late sixties just so I could hear the music firsthand.

Let's put it this way: I had a sitar before George Harrison got his. I wouldn't say I played it as well as he did, though; I think George used it well. The Beatles' "Within You, Without You" is extremely tasteful. He spent a lot of time studying with Ravi Shankar, and it showed. But I can remember going to see Ravi in concert very early on. To show you how far back it was, there were no young people in the audience at all, just a lot of older people from the Indian embassy. This girl I knew was a friend of his, and she took me to see him. She introduced us after the concert and I explained that I had a sitar but I didn't know how to tune it. He was very nice to me and wrote down the tuning on a piece of paper.

But it's really hard for me to say exactly where I got my technique, because it's a combination of a lot of things that were floating around. Sometimes I tell people it's a product of my "CIA connection"—which is shorthand for Celtic, Indian, and Arabic music.

You recorded "Baby Come On Home" during the Led Zeppelin I sessions, but it didn't see the light of day until 1993's Boxed Set 2.
I don't think we finished it—the backing vocals weren't very clever. And at the time we thought everything else was better. Simple as that, really. But don't get me wrong, the track is good and Plant's singing is excellent. It's just that we set such high standards for ourselves.

How did Atlantic react when you presented them with the first album?

They were very keen to get me. I had already worked with one of their producers and visited their offices in America back in 1964, when I met [Atlantic executives and producers] Jerry Wexler and Leiber and Stoller and so on. They were aware of my work with the Yardbirds because they were pretty hip people, so they were very interested. And I made it very clear to them that I wanted to be on Atlantic rather than their rock label, Atco, which had bands like Sonny and Cher and Cream. I didn't want to be lumped in with those people—I wanted to be associated with something more classic. But to get back to your question: Atlantic's reaction was very positive—I mean they signed us, didn't they? And by the time they got the second album, they were ecstatic.

I was looking at some old photos of the band recently, and I noticed you were using an assortment of fairly bizarre amplifiers and guitars around the time of the first album. What were you using before you switched to the Marshall Super Lead/Les Paul combination that most people associate you with?
It was basically whatever we could afford at that time. I didn't really make any money when I was with the Yardbirds, so I was pretty broke in the beginning. I actually had to finance the first Zeppelin album with money I had saved as a session musician. What I had as equipment was very minimal. I had my Telecaster that Jeff Beck gave me, a Harmony acoustic, a bunch of Rickenbacker Transonic cabinets left over from the Yardbirds, and a hodgepodge of amps—Vox and Hiwatts, mostly.

I also had a black Les Paul Custom with a tremolo arm that was stolen during the first eighteen months of Zeppelin. It was lifted at the airport. We were on our way to Canada, and somewhere there was a flight change and it disappeared. It never arrived at the other end.

What are you playing on the first album?
Primarily the Telecaster.

That Tele sported quite a spectacular psychedelic custom paint job.
I painted it myself. Everyone painted guitars back then.

What happened to it?

I still have it, but it's a tragic story. I went on tour with the '59 Les Paul, and when I got back, a friend of mine had "kindly" painted over my paint job. He said, "I've got a present for you." He thought he had done me a favor. As you can guess, I was *real* happy about that. His paint job totally screwed up the sound and the wiring, so only the neck pickup worked. I salvaged the neck and put it on my brown Tele string bender that I used in the Firm. As for the body . . . it will never be seen again!

Did you switch to the Les Paul on the second album because you felt that at some level the Tele wasn't cutting it?

No. If you listen to the first album, the Tele is doing all of that. It was certainly doing the job.

So why did you leave it behind?

When Joe Walsh was trying to sell me his Les Paul, I said, "I'm quite happy with my Telecaster." But as soon as I played it, I fell in love. Not that the Tele isn't user-friendly, but the Les Paul was so gorgeous and easy to play. It just seemed like a good touring guitar.

You produced, but what role did your engineers play in coming up with the sound of the first two albums?

Glyn Johns was the engineer on the first album, and he tried to hustle in on a producer's credit. I said, "No way. I put this band together, I brought them in and directed the whole recording process, I got my own guitar sound—I'll tell you, you haven't got a hope in hell." So then we went to George Chkiantz and Eddie Kramer for the second album and Andy Johns after that. I consciously kept changing engineers because I didn't want people to think that they were responsible for our sound. I wanted people to know it was me.

Did Eddie Kramer have an impact on *Led Zeppelin II*?

Yes, I would say he did, but don't ask me what. It's so hard to remember. Wait, here's a good example. I told him what I wanted to achieve in the middle of

"Whole Lotta Love," and he absolutely helped me get it. We already had a lot of the sounds on tape, including a theremin [an electronic instrument used to create the high siren-like effects] and slide with backward echo, but his knowledge of low-frequency oscillation helped complete the effect. If he hadn't known how to do that, I would have had to try for something else. So in that sense, he was very helpful. Eddie was always very, very good. I got along well with him and, I must say, when I went through all the old recordings for the box sets, all of his work held up very, very well. Excellent!

Zeppelin II **was recorded in many different places.**
It began with rehearsals at my home in Pangbourne, and "Whole Lotta Love" and "What Is and What Should Never Be" were recorded at London's Olympic Studio Number 1 with Chkiantz, who engineered the basic tracks and some guitar overdubs. Those sessions provided the foundation for the rest of the tracks that were recorded and overdubbed at various studios while touring in America. We recorded and overdubbed our way from West Coast to East Coast. Part of the excitement of the album is due to the fact we were completely energized from the live shows and the touring. We did the final mixes with Eddie Kramer at A&R Studios in New York.

What would you say was your greatest achievement as a producer/engineer?
As I mentioned before, miking drums in an ambient way—nobody was doing that. I kept exploring and expanding that approach, to the point that we were actually placing mikes in the hallways, which is how we got the sound on "When the Levee Breaks" [from *Led Zeppelin IV*]. That was purely in the search for ambience and getting the best out of the drums. So it was always better for me to find an engineer who knew exactly what I was talking about. After a while, I didn't have to argue because they knew that I knew what I was talking about.

Speaking of Eddie Kramer, did you ever jam with Jimi Hendrix, who worked so closely with Eddie?
No. And I never saw him play, either. Back in the late sixties, I went right from working with the Yardbirds to touring and recording with Zeppelin,

and that kept me very busy. In the first two years of any band, you just work solidly; if you're going to make an impact, that's what you have to do. We were no different. In fact, we probably worked for three years straight. Anyway, every time I came back from tour and Hendrix was playing somewhere, I would always say to myself, "Oh, I'm just so exhausted, I'll see him next time." I just put it off and, of course, there ultimately never was a "next time." I'm really, really upset with myself for never seeing him. I really wanted to hear him.

As a producer, what did you think of his records?
I thought they were excellent. Oh, yeah. Jimi's drummer, Mitch Mitchell, was also a man inspired. He never played drums like that before or after. He played some incredible stuff!

Although your playing styles were different, you and Jimi were similar in that you both tried to realize these great aural landscapes.
Well, there were a lot of people going in that direction. Look at the Beatles. Here was a band that went from "Please Mr. Postman" to "I Am the Walrus" in a few short years.

What other musicians of that era do you feel had a vision of the future?
Syd Barrett's writing with the early Pink Floyd was inspirational. Nothing sounded like Barrett before Pink Floyd's first album. There were so many ideas and so many positive statements. You can really feel the genius there, and it was tragic that he fell apart. Both he and Jimi Hendrix had a futuristic vision in a sense.

Why do you think records are less dynamic these days than they were in the sixties and seventies?
Well, you had no drum machines in those days. You had to play everything, so there were all these natural crescendos and great ambient sounds to work with. But things like panning and extreme positioning make for a very exciting listening experience. One of my favorite mixes is at the end of "When the

Levee Breaks," when everything starts moving around except for the voice, which remains stationary.

I'll tell you a funny story about that song. Andy Johns did that mix with me, and after we finished it, Glyn, Andy's older brother, walked in. We were really excited and told him, "You've got to listen to this." Glyn listened and just said, "Hmmmph. You'll never be able to cut it. It'll never work." And he walked out. Wrong again, Glyn.

When you were borrowing from classic blues songs on the first two albums, did you ever think it would lead to trouble for you?

You mean getting sued? [In 1987, blues songwriter and bassist Willie Dixon, citing similarities between his "You Need Love" and "Whole Lotta Love," sued Led Zeppelin for plagiarism. The case was settled out of court.] Well, as far as my end of it goes, I always tried to bring something fresh to anything that I used. I always made sure to come up with some variation. In fact, I think in most cases you would never know what the original source could be. Maybe not in every case—but in most cases. So most of the comparisons rest on the lyrics. And Robert was supposed to change the lyrics, and he didn't always do that—which is what brought on most of our grief. They couldn't get us on the guitar parts or the music, but they nailed us on the lyrics.

We did, however, take some liberties, I must say. But never mind: We did pay! Curiously enough, the one time we did try to do the right thing, it blew up in our faces. When we were up at Headley Grange recording *Physical Graffiti*, Ian Stewart [the Rolling Stones' unofficial keyboard player] came by and we started to jam. The jam turned into "Boogie with Stu," which was obviously a variation on "Ooh! My Head" by the late Ritchie Valens, which itself was actually a variation of Little Richard's "Ooh! My Soul." What we tried to do was give Ritchie's mother credit, because we heard she never received any royalties from any of her son's hits, and Robert did lean on that lyric a bit. So what happens? They tried to sue us for all of the song! We had to say bugger off. We couldn't believe it.

So, anyway, if there is any plagiarism, just blame Robert. [*laughs*] But seriously, bluesmen borrowed from each other constantly, and it's the same with jazz; it's even happened to us.

You've often mentioned your folk and rockabilly influences in the past, but what were some of your favorite blues records and guitarists?
I had lots of favorites. Otis Rush was important—"So Many Roads" sent shivers up my spine. There were a number of albums that everybody got tuned in to in the early days. There was one in particular called, I think, *American Folk Blues Festival*, which featured Buddy Guy—he just astounded everybody. Then, of course, there was B. B. King's *Live at the Regal*. The first time I heard any of these people—Freddie King, Elmore James—it just knocked me flat.

Now that I've said all that, I'm missing one important person—Hubert Sumlin. I loved Hubert Sumlin. And what a complement he was to Howlin' Wolf's voice. He always played the right thing at the right time. Perfect. [Howlin' Wolf's "Killing Floor" and "How Many More Years" were sources of inspiration for Zep's "The Lemon Song" and "How Many More Times," respectively.]

Moving on to *Led Zeppelin II*, what was the impetus for the unaccompanied solo in the middle of "Heartbreaker"?
I just fancied doing it. I was always trying to do something different, or something that no one else had thought of. But the interesting thing about that solo is that it was recorded after we had already finished "Heartbreaker"—it was an afterthought. That whole section was recorded in a different studio and was sort of slotted in the middle. If you notice, the whole sound of the guitar is different.

I've actually noticed that the tuning of the guitar was slightly higher.
The pitch is off as well? I didn't know that! [*laughs*]

Was that solo composed?
No, it was made up on the spot. I think that was one of the first things I played through a Marshall as well.

How did you achieve your sound on "Whole Lotta Love"? It's a sound that is incredibly hard to duplicate.

I've talked about distant miking on drums, but I also used it on my guitar sounds. Distance makes depth, which in turn gives you a fatter guitar sound. The first album was done totally with a Vox Super Beatle, so with just that and a Telecaster, a wah-wah, and a boost pedal, you can create a great variety of sounds. I used a depressed wah pedal on the solo. I did the same thing on "Communication Breakdown." It gets you a really raucous sound that just slices through everything.

The band recorded a number of sessions for BBC Radio between 1969 and 1971, and you officially released them years later in November of 1997. When you dusted them off, what did you find most surprising about them?

Historically, they're good because those sessions are exactly what they are: a snapshot of a band trying to make their way through a very busy schedule. I've heard the sessions several times throughout the years, so they weren't completely fresh to my ears. But what I find most exciting about them is comparing the different versions of the same songs. It's interesting to hear how a song like "Communication Breakdown," which appears three times, evolved from performance to performance. It's like looking at a diary. It's not the best Led Zeppelin, and it's not the worst. It's what it was that night. Which, in most cases, was very good.

The *BBC Sessions* show in graphic detail just how organic the group was. Led Zeppelin was a band that would change things around substantially each time it played. The two performances of "You Shook Me" are particularly good examples of what I'm talking about. The version that opens up the album is not shabby by any standards, but it doesn't compare to the second version, which was recorded only a few months later. The interplay between Robert and me had grown in leaps and bounds. That kind of thing was a subtle indication of how the band was really beginning to gel. We were becoming tighter and tighter, to the point of telepathy.

I mean, compare our sessions to, say, the BBC recordings of the Beatles. I bet you a cent to a dollar, if they have two or three versions of "Love Me Do" or whatever, they'll all be identical. That was the difference between our contemporaries and us: Led Zeppelin was really moving the music all the time.

You edited some of the performances. Give me an example of the type of work you had to do.

I didn't have to do much. The biggest problem was editing down the ninety-six-minute Paris Theater performance to the length of one CD, which is eighty minutes. Most of the editing was done on one song in particular—the "Whole Lotta Love" medley, which originally ran over twenty minutes. It included the band playing bits of "Let That Boy Boogie Woogie," "Fixin' to Die," "That's Alright Mama," and "Mess o' Blues," all of which were left in; and "Honey Bee," "The Lemon Song," and "For What It's Worth," which were edited out.

It was actually amazing the kind of editing we were able to do with Pro Tools software—I was able to move things around quite a bit on the medley without disturbing the groove. For example, half of one of my solos is edited into the second half of another solo and you'd never know! It's the kind of thing I just love doing. I really enjoy being given a problem that looks insurmountable and finding a creative solution.

Are there any gaffes that you left in?

There's a moment in the second version of "You Shook Me" that's funny and intense. The guitar comes in really loud, and you can tell that the engineer was caught by surprise because he panics and whacks the fader down. We left it because it's a very real moment. So just in case anybody thinks it's my fault, don't blame me, I wasn't the engineer! [*laughs*]

But it's important to understand that, at the time, that kind of mistake wasn't a big deal. We never dreamed that there would be these things called CDs or that people would still be interested in any of the sessions twenty-five years later. It was purely for broadcast, and perhaps a repeat.

I noticed that some of the tracks have simple overdubs—maybe an added rhythm guitar behind your solos, or an extra harmony on one of the versions of

"Communication Breakdown." Were these things you insisted on, or were they common practice?

I think at the time the BBC just wanted the best show possible. They realized that bands were often trying to re-create something that they had created on an eight-track machine, so doing an overdub here or there was fair game. At least one overdub, anyway, which is all we ever did.

What kind of shape were the original BBC tapes in?

That's a story in itself. When we first started entertaining the idea of releasing the sessions, we asked the BBC to send over copies of what they had. They sent Jon Astley, who assisted me in the remastering process, a DAT copy of the recording. Because they didn't send us the original quarter-inch masters, we just assumed they must have transferred everything to digital and wiped the original tape. The plan was that we were going to remaster and edit from the DAT.

The interesting thing was, when I asked the BBC to send over the copies, I specifically requested a cassette tape, which I thought would be duped from the DAT. However, when we sat down to start remastering and Jon played me the digital tape, I thought, My God, this sounds terrible. My mind might be playing tricks, but I think my cassette actually sounds much better than the DAT copy.

So I pulled out my cassette and, sure enough, my tape sounded much better. I then concluded that the cassette must have been duped from a different source—perhaps the original quarter-inch tape. So, following my hunch, we actually found the master tapes in their vaults. It took some searching, but it was worth the effort.

I was under the impression that the BBC regularly wiped their tapes.

I was under the same impression. Actually, I'm somewhat amazed our sessions weren't erased ages ago. You'd imagine they wouldn't have taken rock so seriously.

It's no secret that the Led Zeppelin BBC sessions are among the most bootlegged performances in rock history. They're kind of a hard-rock version of Bob Dylan's

Basement Tapes. To put it another way: If you're a serious Zeppelin fan, you've either heard or own some version of them. What's your view of bootlegging?

It depends. If it's someone with a microphone at a gig, that's one thing. They paid for a ticket, so it's fair game. But things that are stolen out of the studio—works in progress, rehearsal tapes, and things like that—are quite another. I'm totally against that. It's theft. It's like someone stealing your personal journal and printing it.

Regarding the BBC sessions, it doesn't bother me if they've been bootlegged extensively because, whatever version people own, it won't be from the original source like our own version. Secondly, not everybody out there buys bootlegs, and the sort of fans that buy bootlegs will want this one for the packaging—and to see how I've edited the performances. I can't really lose. There's *no way* I'll lose! [*laughs*]

The idea of bands regularly performing live on radio is such a foreign concept in America. Could you give us some background as to the situation with British radio in the sixties?

Actually, there was a tradition of live radio in America that extended well into the fifties, because I know that Elvis Presley used to play on local radio stations. But it probably died out in the U.S. as it got easier for smaller stations to just spin records.

But in England, live radio never died. The BBC, which controlled everything, continued to produce live dramas, quiz shows, discussion programs, classical music broadcasts, and so on. And as rock started becoming part of the culture, it just became part of the BBC's mix.

In keeping with BBC practice, your early sessions were very short. What were the circumstances that led up to your hour-long performance at the Playhouse Theater in 1969 and the ninety-minute show at the Paris Cinema in 1971?

When we first started participating in the sessions, we were only given enough time to play two or three songs. We started complaining—as did many other groups—that we couldn't represent the band properly within those time constraints. I guess they finally took us seriously, because they allowed us to pilot an "in concert" show that would allow us to play a complete, hour-long set.

Our pilot was so successful that it soon became the standard format. That was kind of an important first and, I guess, shows the kind of clout we had in those days.

You seem to have played "Communication Breakdown" in the majority of your BBC shows. Any particular reason?

I think we felt at the time that "Communication Breakdown" and "Dazed and Confused" in particular were most representative of what the band was all about.

Legend has it that the arrangements of three songs featured on the BBC sessions—"The Girl I Love" by Sleepy John Estes, "Traveling Riverside Blues" by Robert Johnson, and "Something Else" by Eddie Cochran—were made up by the band almost right on the spot. Considering how important your radio appearances were, it was pretty ballsy to improvise in such a manner.

Yeah, it was. Fortunately, with a band like Led Zeppelin, it was no problem to do something like that. Everybody was so on top of it and *alive*—individually and collectively. I'd say, "I've got a riff." I'd show it to everybody. We'd bang through it once, Robert would sing something over it, and then we'd record it. It was that simple.

Speaking of "Traveling Riverside Blues," one of the more striking differences between the two *BBC Sessions* CDs is that by 1971 you had pretty much stopped covering traditional blues songs.

Well, we started writing our own blues songs, didn't we? After the first album, I was really conscious of the fact that we had to start carving our own identity. I felt a particular pressure to make my unique personal contribution.

In the early days, I was quite happy to borrow from Otis Rush for "I Can't Quit You." It was a pleasure. But after a while I started realizing that that wasn't what I should be doing. I felt I had to start developing my own thing and almost stop listening to anybody. And I think I succeeded.

A good example is "Nobody's Fault but Mine," from *Presence*. Robert came in one day and suggested we cover it, but the arrangement I came up with was nothing to do with the [Blind Willie Johnson] original. Robert may have wanted to go for the original blues lyrics, but everything else was a totally different kettle of fish. Actually, I only recently discovered that "The Girl I Love" had something to do with Sleepy John Estes!

Do you think the BBC shows helped break the band in the UK? It has often been noted that success came much more slowly for you in England than in the States.
That's a myth. It's just not true. When we first started playing in England, it might have been a little difficult because the first album wasn't released yet, and people hadn't heard the band. And that's hard—it's hard to start out that way, even if you're really, really good and knocking people dead in their tracks. But I had a big reputation from the Yardbirds, and people were really keen to see what I was doing. And as soon as they saw us, the word of mouth started spreading and we became popular pretty quickly. If you gauged popularity by concerts, the attendance was equal on both sides of the ocean.

By the time we got to the States, people were already becoming familiar with the record, and that made it easier. It also didn't hurt that American FM radio was supportive. But in both countries, Led Zeppelin was a massive word-of-mouth thing, really. It was just a general ambience that crept in everywhere. [*laughs*]

You spent a lot of time in America in the early days. Did you find that at all intimidating, given how comparatively small the UK is?
The Yardbirds were very popular there. But above and beyond that, I could see the potential there. The audiences were much bigger, and there were naturally more places to play and more opportunities for the band on every level.

I would imagine that touring in the late sixties was primitive compared to how it's done these days.
People ask me whether it was difficult back then, but it never really seemed

hard because that's what we were used to. There were no tour buses back then—you just hired cars and flew on commercial airliners. That's how it was, and that's what we were used to.

If you're fortunate enough to make a record, and you've got your foot in the door, the most important thing is to get as much exposure as humanly possible. Albums were things that were fitted in between tours. In fact, as you well know, all of *Led Zeppelin II* was recorded at various studios while we were on the road and mixed at the end of a tour. When I look back, I have to admit our stamina was quite phenomenal.

I was struck by how clean your sound is on the BBC sessions. What is your philosophy regarding volume?

The answer is, I turn up pretty high, but I vary my pick attack—I don't play hard all the time. I find that this approach helps me get more tonal and dynamic variation, especially when I'm playing close to the bridge or close to the neck. Then you have the power if you really want to hit it hard. If you go hard all the time, you just won't get the difference in tone.

How did the four members of Zeppelin interact on a personal level? Was everything as smooth internally as it appeared to be?

I think the atmosphere in Led Zeppelin was always an encouraging one. We all wanted to see the music get better. And part of the reason things ran smoothly is that I had the last decision on everything. I was the producer, so there weren't going to be any fights.

The atmosphere was always very professional. I was meticulous with my studio notes, and everybody knew that they would get proper credit, so everything was fine.

Another key: We all lived in different parts of the country, so when we came off the road we didn't really see each other. I think that helped. We really only socialized when we were in the studio or out on the road. We all really came to value our family lives, especially after being on the road so much, which is how it should be. It helped create a balance in our lives. Our families helped keep us sane.

MUSICAL INTERLUDE

———————

A CONVERSATION WITH
JOHN PAUL JONES

Bassist/keyboardist John Paul Jones reflects
on his working relationship with Jimmy Page
and his time in Led Zeppelin.

———————

JOHN PAUL JONES was the stoic eye of the Led Zeppelin hurricane—the seasoned pro who valued a "really solid foundation" above all else. While Jimmy Page and Robert Plant exploded onstage, Jones stayed in the background, content to focus his energy on anchoring one of rock's most dynamic rhythm sections.

"I used to enjoy locking into John Bonham's drums very tightly—I suppose it was my session background," Jones says. "A good session was one where the rhythm section really locked together. In Led Zeppelin, I would listen to the bass drum and be very careful not to cross it or diminish its effectiveness. I really wanted the drums and bass to be as one unit—that's what drove the band along. It was important to be rock solid so Jimmy and Robert could be more free to improvise and experiment."

"Robert always used to say that onstage I should stand much nearer to the front—get some light on me and all that, from the visual angle," he says, chuckling. "And I would try. I would start at that front and I would just move backwards and backwards. I would always end up in my favorite position, which was as close to the bass drum as possible."

Jones, it seems, has always been pretty comfortable working behind the scenes. Born John Baldwin in 1946, the bassist began his professional career in the early sixties, touring with various bands in England. Like Page, Jones eventually drifted into studio work, playing his 1961 Fender Jazz bass on hundreds of sessions from 1962 to 1968.

"I was always in demand because I was one of the only bassists in England that knew how to play a Motown feel convincingly in those days," Jones explains. But when it quickly became evident that he was capable of much more than copping a sweet soul groove, Jones, who was still in his teens, was booked to perform on everything from jingles to Herman's Hermits singles.

In addition to playing bass, Jones also began building a reputation as a talented arranger, writing charts for such psychedelic nuggets as Donovan's "Mellow Yellow" and the Rolling Stones' "She's a Rainbow." But by 1968 he was, in a word, "fried."

He vividly recalls: "Being a session arranger is literally a twenty-four-hour job—composing the individual cores for horns and strings the night before, handing them out the next day, and knocking out the finished product. I was arranging fifty or sixty things a month, and it was starting to kill me."

Enter Jimmy Page.

Page, who worked with Jones in the early sixties when both were thriving session prodigies, had one year earlier bid farewell to the grind of the studio to join the Yardbirds. After the Yardbirds split, Jones heard through the grapevine that Page was starting a new band. In desperate need of a break from his studio work, Jones rang Page up to see if he needed a bassist. Page needed little convincing. As the guitarist remembered in a 1969 Zeppelin press release, "I was working on the Donovan album *Hurdy Gurdy Man* with John, who composed some of the arrangements. He asked if I could use a bass guitarist in Led Zeppelin. I knew that he was an incredible musician. He didn't need me for a job, but he felt the need to express himself and figured we could do that together."

When you're right, you're right.

Led Zeppelin: (from left) John Bonham, Robert Plant, Page, John Paul Jones, 1968 (© *Getty Images*)

It's no secret that Led Zeppelin broke in the United States before doing so in England. Why was it easier to break in America?

JOHN PAUL JONES We toured there first—nonstop. And American radio, which was completely different from British radio, allowed us to penetrate the country a lot quicker. FM radio was just starting to get real strong in the U.S., and they really supported the band. We in turn fanned the flames by visiting every shed, shack, and chicken run that had an FM aerial sticking out of it. FM radio is something you take for granted now, but it was very new and exciting back then.

Before joining Zeppelin, you had a thriving career as a session player and arranger. In the early days of the band, when you were playing one tour after another, did you ever find yourself thinking, *What* have I gotten myself into?

No, never, because arranging was crazy, too. And after a while it wasn't much fun, either. I joined the band to escape that life and everything attached to it.

Was it a jolt, exchanging the relatively sedate life of a studio musician for that of a rocker on the road?
It wasn't foreign to me. I had been in a group before I started doing sessions, so I was already used to the rigors of touring. I had played in this jazz band called the Tony Meehan–Jet Harris Combo, which somehow ended up on the road with an assortment of English and American rock acts like Del Shannon and the Four Seasons. I had experienced the cars, the travel, and adulation, and it wasn't that different from touring with Zeppelin.

But it had to be somewhat different. Led Zeppelin audiences were most unlike Del Shannon's, for example.
That's true: the audiences were completely different. Instead of screaming teenyboppers, people were very interested in what Zeppelin was playing. They actually listened.

Maybe they had no choice because the band was so loud!
[*laughs*] We really weren't that loud in the beginning. But the audience was much more open, especially in the early days. Later on, when we were playing stadiums, I found the shows became more like an event than an opportunity to listen to the band. And the bigger and bigger the venue, the less interesting the band became, musically.

As the stages got bigger, it became more difficult for the band to communicate and relate to the audience. I remember having the feeling that I could play anything and it really wouldn't matter. Not that I did, mind you, but there was that feeling. We were playing football stadiums in the days before large video screens, so all we were, to most of the audiences, was a speck in the distance. The majority couldn't see us, and I'm convinced they also couldn't hear us, either. But despite all that, the band always worked very hard. We were always professional.

Professional **is a key word in describing the band. You really sustained a high level of creativity and discipline over a remarkably long period of time.**

I think it's even more remarkable in light of the fact that we had no restraints over what we recorded. As we never had anyone from Atlantic looking over our shoulders, it would have been incredibly easy to coast. But we never did.

Do you think this professionalism was an outgrowth of the approach you and Jimmy developed during your session days?

Could be. If you didn't deliver, you didn't get called back.

Much has been made of Zeppelin's eclecticism. Could this also be a stepchild of the studio, where you frequently were required to play a wide assortment of musical styles?

We did have to play whatever was thrown in front of us, from country to reggae to middle of the road, sometimes all in the same day. So, yes, we did have the technical ability to do all of those things. Add this to the fact that we all had completely different tastes in music, and you have a fairly clear understanding of our eclecticism.

I think the problem with most modern bands is that all the band members listen to the same music, which makes for a very one-dimensional sound. We never listened to the same music.

Logic suggests that a band in which everyone goes in different directions is a band ripe for conflict.

No, not at all. We considered it valuable. I've always maintained that Zeppelin was the spaces between us. Bonzo was into soul music and Motown ballads; I was into jazz and classical music; Jimmy was into rockabilly, blues, and folk; and Robert was into blues and Elvis Presley. None of us had the same record collection. Nobody on the outside of the band could understand this.

But it was very clear to me why it worked. We all loved music and liked learning about new things. Each of our individual record collections was interesting to everyone else. I'd go over to Robert's house or Jimmy's house and hear some blues I would otherwise never have been exposed to. For instance, I'd never actually heard Robert Johnson before joining Led Zeppelin.

It's surprising that you weren't a huge blues fan, considering how blues-based early Zeppelin was.
I was really into jazz, though, and jazz came from the blues, so it felt very natural to me.

While Zeppelin improvised, it never seemed . . .
. . . gratuitous?

Yes. Did the band learn from the mistakes of Cream and the Grateful Dead, who often fell victim to their own improvisational self-indulgence?
The only rock I listened to in the sixties was Jimi Hendrix—the rest of the time I was focusing on jazz and soul. I never listened to Cream or the Dead. I wouldn't know what they did. I don't think anybody else in the band listened to them either. Maybe Jimmy was more aware. We just relied on our own musical sensibilities to edit our music, in improvisation or writing or anything else. If something goes on too long, just stop it, for God's sake!

Jimmy had his vision of how he wanted the band to sound going into the project, and I certainly knew how to play. There were definitely lines and directions we wanted to follow, and we knew how to accomplish them. But the odd thing is, we didn't really have to discuss our ideas. You could hear what was happening, and you instinctively knew what should happen.

The three-hour Zeppelin show was legendary. How did it evolve?
We simply could not control ourselves. You have to be interested in what you are doing before you can interest other people. And we really liked to play. Our songs were structured so they could take off at any given moment, and if we were having a hot night, invariably they would.

Believe it or not, we started as a forty-five-minute act, which I think we were able to contain ourselves to twice. Then it became a game of "Let's play anything anybody knows twelve bars of." I know it sounds like a recipe for disaster, but we knew we could always pull it off. We knew how to make things sound good because we were all experienced musicians.

The other secret was that we were a selfless band. Everybody was in tune with everybody else, and everybody always listened to everybody else. That was the key. The feeling was never "What am I doing?" It was always "What is the *band* doing?"

Zeppelin was one of the first bands to break away from the more casual hippie look of the sixties and adopt a more glamorous image.
I think that at the time we broke, there were a lot of hippie bands who just wandered around onstage, looking at each other between songs and just sort of wondering what to do, wondering what key they were in . . . just wondering. We wanted to eliminate that. We wanted to hit people hard and put on a show, which included dressing a certain way. We wanted to look good, sound good, and play good. And at the time, not many people were putting as much conscious effort into every aspect of live performance as we did.

The hippie approach was generally "Come as you are and do what you feel like," and we completely electrified audiences because our act was so focused on every level. Again, it was our professional attitude. People were paying a lot of money to come and see us, so you wanted to sound like something and look like something. And people appreciated it when you put in the effort. It doesn't detract from the musicianship and it adds to the show, so why not?

How did the band interact on a personal level? It seemed like a very enclosed world, which added to the band's mystique.
The band was very close, which fostered a feeling of "us against them." We knew what to do and we knew how to do it; beyond that, we didn't want to be bothered. We had the same attitude while making our records. Our manager, Peter Grant, did a very good job of keeping everyone away from us, which allowed us to focus on our work.

There really isn't any mystery. We always got on well. We never socialized when we weren't touring, but we were always pleased to see one another. We never went through that bickering stage that you hear about other bands. We shared a professional state of mind. We were always reliable. You could probably count the number of Zeppelin gigs that were canceled on one hand. We were always where we were supposed to be.

PAGE AND SINGER ROBERT PLANT HEAD FOR THE
WELSH COUNTRYSIDE, SYNTHESIZE FOLK, BLUES,
AND ROCK, AND RECORD *LED ZEPPELIN III*.

Page at his Thames boathouse in Pangbourne, 1970 *(© Mirrorpix)*

"FUCK THE SIXTIES! WE'RE GOING TO CHART THE NEW DECADE . . ."

WHILE *LED ZEPPELIN IV, Physical Graffiti,* and even *Led Zeppelin II* get the lion's share of attention, one could argue that Zeppelin's third album, released on October 5, 1970, is equally important. Often marginalized as "the acoustic album," *Led Zeppelin III* was much more than that: It represented a sophisticated leap in synthesizing the folk, blues, and rock elements found on the band's first two albums into what one thinks of as "the Led Zeppelin style." Otherworldly songs like "Friends," "Immigrant Song," and "Celebration Day" were such a departure that it took the critics and public another album before they could completely comprehend the band's daring new synthesis.

"Led Zeppelin was definitely growing, there's no doubt about that," Page says. "Where many of our contemporaries were narrowing their perspective, we were really being expansive. I was maturing as a composer and player, and there were many kinds of music that I found stimulating, and with this wonderful group I had the chance to be really adventurous. It was the same for everybody in the band. Because with the high level of musicianship and creativity of the four members, we were really able to approach anything—*attack* anything."

The story of *Led Zeppelin III* starts, appropriately, at the dawn of the seventies, an exciting new decade during which the band would dominate the rock world. After completing their fifth tour of America, Page, Plant, Jones, and Bonham returned to England for a well-deserved break. In a scant year and a half, Led Zeppelin had played almost two hundred shows, recorded two best-selling albums, and seen their concert performance fees climb from $1,500 to an astronomical $100,000 a show. Their hard work and discipline had paid off, but it was time to rest and recharge their creative batteries.

It was Plant's idea that he and Page might benefit from a vacation in the spring of 1970. Plant recalled an eighteenth-century cottage called Bron-Yr-Aur, in Wales's Cambrian Mountains, that he had visited as a child with his family. He decided that a return visit was in order and extended an invitation to Page. The two men and their families packed up guitars and supplies and headed off for a retreat.

Page was accompanied by Charlotte Martin, a striking French model he met on his twenty-sixth birthday in 1970. Introduced to Jimmy by the Who's Roger Daltrey after a Zeppelin performance at London's Royal Albert Hall, Martin would be Page's constant companion throughout the next decade, giving birth to their daughter, Scarlet Lilith Eleida Page, on March 24, 1971. (The January Albert Hall show was notable for other reasons—it was filmed and eventually released on the *Led Zeppelin* DVD in 2003. See chapter seven for more details.)

The absence of electricity in the cottage guaranteed that any music created on their trip would be acoustic. It was just as well—the quiet came as a welcome relief for the musicians, who had just spent months playing music at top volumes.

"When Robert and I went to Bron-Yr-Aur, it wasn't like it was, 'Let's go down to Wales and write,' " Page says. "The original plan was to just go there, hang out, and really appreciate the countryside. The only song we really finished up there was 'That's the Way,' but being in the country set a tone, and it established a standard of traveling for inspiration."

While *Led Zeppelin II* was primarily a snapshot of a touring band caught in the heat of battle—ferocious and filled with testosterone—*Led Zeppelin III* would introduce a new sensitivity to the band's overall sound. Compositions like the delicate "That's the Way," the East Indian–influenced "Friends," and the jaunty country hoedown "Bron-Y-Aur Stomp" added considerable new depth to the Zeppelin oeuvre and opened new avenues for future albums.

Led Zeppelin III will be forever known as Zeppelin's acoustic album, but that is something of a misnomer. For every gentle yin, there is plenty of ballsy yang. The relentless attack of "The Immigrant Song," the pummeling thunder of "Out on the Tiles," and the hypnotic, backward-sounding groove of "Celebration Day" proved that Page still knew how to rip rock's most imaginative and aggressive riffs.

CONVERSATION

Q:

WHERE DID *LED ZEPPELIN III*
BEGIN?

The first two things I had for the third album were "Immigrant Song" and "Friends," which wasn't a bad place to start. "Immigrant Song" had a great driving riff, which spoke for itself. "Friends," on the other hand, was more exotic—it had a North African or Indian flavor. I remember I was playing around with this open-C tuning [guitar strings, low to high: C A C G C E], but before I had written anything, I had a massive argument with my ex-wife. I went out on a balcony in my house and suddenly the whole song spilled out, just like that. Considering the song's origins, it's ironic that it ended up being called "Friends." [*laughs*]

In many ways, those songs were two sides of the coin for the third album—the electric and the acoustic. "Since I've Been Loving You" was also around, but it was a work in progress. We'd been playing it live, but hadn't been able to capture it in the studio.

While "Immigrant Song" is built around a very straightforward, pile-driving riff, it's the subtle variations in it that make it more than just another hard-rock song. For example, toward the very end of the song, instead of playing a straight G minor for the accents, you play this very astringent inversion of that chord that really adds some bite. Where did that come from?

It's a block chord that people never get right. It pulls the whole tension of the piece into another area or another dimension just for that moment . . . and a bit of backwards echo makes it a bit more complete. It's putting all these elements together that makes the music have depth.

I have to say that Robert's input on that song was also absolutely magnificent. His sort of "Bali Ha'i" [a song from the classic Broadway musical *South Pacific*] melody line was really inspired and completely spontaneous. I can remember working on "Immigrant Song" and all the pieces coming together. John Bonham and I playing the riff, putting in the E to A "Rumble" chords, and Robert singing his wonderful melodies . . .

So to answer your question, where did that unusual G chord come from? I didn't have that chord when I started writing the "Immigrant Song," but it suddenly appeared while we were working together, putting on a massive brake to this machine. You know those old brakes where you clutch them and it just pulls back out again, pulling it back in—that's how I see the function of that chord.

I don't think many typical rock players would come up with that.
No.

They wouldn't really have that vocabulary . . .
. . . or perhaps the cheek or audacity of inserting that chord. It's like, "Oh, really? And what is that?" [*laughs*] Not only was it audacious, but it's a chord that nobody could work out, which is even better.

A secret chord!
Well, yeah. See if they can spot that!

You mentioned Link Wray's "Rumble" within the context of "Immigrant Song." I never would have put that together.
Well it's not, literally. I just think of those slashing E and A chords after the F-sharp octave riff as having the feel and sound of the way Link plays the chords in "Rumble." It's something of that attitude.

So you had "Friends," "Immigrant Song," and "Since I've Been Loving You" written. Where does your sojourn to Bron-Yr-Aur come into the picture?
Well, for the first two years of Led Zeppelin, we were solid at it, and before that I was on the road with the Yardbirds. The pace I was moving at was really phenomenal. When I look back at what we actually did in 1969 alone, it's absolutely mind-boggling.

In early 1970, after our fifth U.S. tour, we took a break. I wouldn't even call it a break—it was just a few days off. It felt like a really substantial break, but if you really look at it, it was just a couple of weeks—hardly any time at all. But Robert and I managed to go to Bron-Yr-Aur in Wales to get a big injection of the countryside. We needed to get away because we had been living a real gritty urban existence.

Were the acoustic elements of *III* influenced by other folk-rock bands of the era like the Band or Fairport Convention?

We liked those bands, but we didn't pay that much attention to what other people were doing or how we fitted in. It was just go up and rock it, and that's what it was.

I remember an absurd press comment comparing us to Crosby, Stills and Nash because of the acoustic elements on the third album. I thought, That's absolutely pathetic, because acoustic guitars were all over the first two albums. It was always there, it was right at the core of everything. It was always meant to be there. The third album was just another evolution. It was different from the second album as the second album was different from the first.

As you said earlier, you were really tearing it up in the sixties. Did the mellower sounds on the third album represent taking a breather from the craziness of that decade?

Well, I didn't really think that. I was into the seventies. Our attitude was, "Fuck the sixties, we're going to chart the new decade!" We were on a mission.

Do you think some of the initial negative reaction to Zeppelin was because the critics looked at the band through a sixties aesthetic, and you had already moved on to sort of the next decade—something that was distinctly the seventies?

Yeah, we were so far ahead that it was very difficult for reviewers to know what the hell we were doing. They couldn't relate to it. Very rarely could they get the plot of what was going on.

In retrospect, your agenda was clear: Led Zeppelin was taking the existing ideas found in traditional blues, folk, and rock and moving them into the future. *Led Zeppelin III* was a substantial leap in that direction.

Okay, okay, well, there it is, then. There was a lot of blues on the first album, but we would have never ventured to play something as unusual or sophisticated as "Since I've Been Loving You." It's another example of our collective energy sparking each other to new heights.

"Since I've Been Loving You" starts as a standard minor blues and slowly unfolds until it touches on just about every chord in the key of C minor in a very natural but dramatic way.

Yeah, right near the end you'll notice it goes to a C7 at one point as opposed to C minor 7.

Can you explain how it evolved? It actually made an appearance at your Albert Hall show in early 1970, so you were working on it before you recorded _III_.

Yes, we played it as part of the Albert Hall set you hear on the _Led Zeppelin_ DVD. The problem with it was that the keyboard didn't get recorded, so there's only the guitar, the drums, and the voice, which is really unfortunate—otherwise we'd had a good version. That's way before we started putting the third album together.

It was a tricky number to record. It was hard to capture the exact dynamics and the overall tension of it, and it was crucial to get the rise and fall of it. We had attempted to record it before and it didn't come off, so we recorded something else instead. There's no point in laboring a song like that, let me tell you; it's either going to happen or it's not. Later we took another crack at it and that worked.

It starts out so skeletal. The chords are only suggested. Then it unfurls into these great crescendos.

To play a blues in C minor is not necessarily that difficult a thing, but our approach was pretty unique. John Paul Jones was definitely integral to creating some of the chordal movement. The people reviewing the album when it first came out literally didn't understand what they were hearing. We all do now, but at the time it was just too much for them to be able to work out the significance.

There are musicians and critics who judge folk and blues music strictly on how authentically the player can reproduce an earlier era. But there's a certain futility to that pursuit. You're never going to completely replicate the music of Muddy Waters or Buddy Guy, so you might as well go someplace else.

That's right. The original Fleetwood Mac with Peter Green, for example, performed the music of people like Elmore James really well. They were right on it. And Peter had such a beautiful touch on things like "Stop Messing Around"—it's just fabulous in the vein of B. B. King.

But with "Since I've Been Loving You," we were setting the scene of something that was yet to come. It was meant to push the envelope. We were playing in the spirit of blues, but trying to take it into new dimensions dictated by the mass consciousness of the four players involved.

The same thing goes for the folk stuff as well. It's sort of, "Well, this is how it was done in the past, but it now has to move." It's got to keep moving, moving. There's no point in looking back. You've got to keep moving onwards. Another factor was that my playing was also improving, and it was developing around the band. I didn't play any of this stuff when I was doing studio work or even in the Yardbirds. I was just inspired with this energy that we had collectively. I don't think there was any way to look backwards.

You had recorded two albums and toured for a solid year and a half with the band. Was the growth due to the fact that you were coming to grips with what each person could do?
No, it was more about understanding the collective and just how far you could push it.

In some ways Zeppelin does remind me of how Muddy Waters and his electric band evolved . . .
I hope so . . . [*laughs*]

Muddy was originally an acoustic country-blues player. But when he left Mississippi for Chicago, he understood his traditional approach to blues wasn't going to work in an urban setting. So he found other musicians with real character, they began amplifying their instruments and took the blues somewhere else. Muddy found a sympathetic unit, and together they pushed the blues into the future.
When Muddy went electric and started creating things like "Standing Around Crying," he really shifted the world. He moved it. That's exactly

what we were after. I'm talking about that sort of tension, that thing, that atmosphere you could cut with a knife. Muddy shakes me up and makes my hair stand on end when I really listen to his stuff. The thing about him is he had all these unique players in his band and he let them shine. Little Walter's harmonica playing on those Chess records is something you almost need to witness to believe.

What do you remember about the actual sessions that resulted in the epic performance of "Since I've Been Loving You"? It features one of your very best solos.
We had maybe two attempts at recording it at two different locations. The intro changed subtly every time we'd play it—certainly from one studio to another—but it had that characteristic first opening. That's quite a traditional way to open up a blues, on those first few notes, isn't it? But beyond that, it was more of a vehicle to push the envelope than playing to please the blues purists.

I wasn't interested in performing a note-for-note rendition to prove to everyone I could play in a certain style. I always approached the blues with a rock and roll rhythm to the phrasing, so whatever I would play was already going to be different. I was exploring a different avenue. I wanted to lock into the overall ambience and atmosphere of the song and what it was conveying. Because after hearing this wonderful construction where everyone is playing so beautifully together and making their own statements—big statements, massive statements, accents and phrases, locking into it and swooning—I had to deliver a solo that would live up to this incredible buildup. It was like getting ready for a hundred-yard dash or something. Just vibing up for it, psyching myself up and coming up with some idea of how to get the solo off, and then . . . go!

It's exactly how I felt before I played my third pass at the "Stairway To Heaven" solo [the one that appears on *Led Zeppelin IV*]. I approached it in exactly the same way. A solo is like a meditation on the song. You find a piece of filigree and then try to play something in total empathy with everything else that's going on.

You can get quite spiritual about soloing. It's almost like channeling.

It's not there one moment, but then all of a sudden it is. I'm sure anyone who's creative has that moment. That point where it just sparks. One minute it wasn't there and the next minute it is, and you know it's positive and constructive.

That's what all musicians look for—finding that moment. As great as "Since I've Been Loving You" is, there is one problem with it.

Yes, there is an awfully squeaky bass-drum pedal on the recording. It sounds louder and louder every time I hear it! That was something that was obviously sadly overlooked at the time.

Led Zeppelin III **is a futuristic folk album in more ways than one. The acoustic songs like "Gallows Pole" are one aspect, but even the heavier electric songs have folk elements. "Immigrant Song" is like an ancient ode, "Celebration Day" is like a bottleneck blues gone berserk, and from what I understand, "Out on the Tiles" started out as a drinking song, which is as folky as it gets.**

Yes, John Bonham had quite a bit to do with "Out on the Tiles." I wrote the opening descending riff, but the guitar part behind the vocal was based on a song he used to sing that went something like: "Out on the tiles, I've had a pint of bitter / and I'm feeling better 'cause I'm out on the tiles." You know what "out on the tiles" means? It's slang for hitting the bars, and a bitter is a sort of dark pale ale. Robert's lyrics took it in a little different direction, but the vibe is still there.

If you think about it, the guitar riff is almost like a fiddle part played three octaves down—it has that sixteenth-note bounce.

Well, that's interesting, yeah—I never thought of it like that, but I suppose it is like a fiddle song.

What did you think about other bands who were modernizing folk and blues, like Fairport Convention or the Byrds? Did you feel any sort of kindred spirit?

I liked all of those people that you've mentioned. But I don't think you'd ever confuse Led Zeppelin with Fairport Convention or the Incredible String

Band. I think they were coming from a much more traditional place, and I was coming from so many different areas. But maybe, really, I was just coming with a rock and roll head! [*laughs*]

Wherever I was coming from, it had lots of dimensions to it. But something like "Friends" really isn't—it isn't traditional music, but I really liked that we could go in that direction and put our own spin on it. At the same time, I don't ever think we lost sight of the fact that we were a rock band.

Well, to that point, "Friends" is an acoustic song, but it is performed with definite rock aggression. That open low C note really rings in the beginning.
Yeah, there's a bit of drama to it. Like I said to you, I wrote it after an argument, and it never lost that urgency.

Actually, the acoustic-guitar sound on all of your albums is always so uniquely "dense." They often have as much presence and hit as hard as your electric sound. Can you offer any insight on how you capture that tone?
Whether you're recording an electric guitar, a drum, or an acoustic guitar, it should all be done with microphones and microphone placement. You shouldn't have to add equalization in the studio if the instruments sound good. But one key to my acoustic sound is that I used an Altair Tube Limiter. I found out about the unit from a chap named Dick Rosmini, who recorded an album entitled *Adventures for 12 String, 6 String and Banjo* in 1964. I'd never heard an acoustic guitar sound quite like that. I bumped into him in the States, and he said the whole secret to his studio sound was the Altair. It turned out to be so good and reliable, we were still using it on *In Through the Out Door* in 1978.

Was the impact of going to Wales exaggerated, or was that really an important moment for you?
Robert and I were the only band members that went out to Wales, but I think it was important insomuch as it functioned as a creative spark. It also sort of set the stage for our later work at Headley Grange [a stone structure, built in 1795, where Led Zeppelin wrote and recorded much of *Led Zeppelin IV* and

Physical Graffiti]. It gave me the idea that we could go somewhere and create a real workshop situation, where we could live it day and night.

Travel has always inspired you.
It's all about the expedition. It's a theme with me that goes back as far as the Yardbirds. When I toured Australia with the Yardbirds, there were two ways to return to the UK. I suggested that we go through India on the way back, but the others opted to go through San Francisco. I thought, I've been to San Francisco, but I may never get another chance to go to India, so I went there on my own and it was really important. There's something amazing about being the only person getting off a plane at three in the morning in a completely foreign land.

In between the first and second leg of Zeppelin's 1977 tour, I decided to go to Cairo because I really wanted to hear the music down there. There's a great picture of me somewhere by the Sphinx. It's all part of my musical history.

In 1972 you took a rather important sojourn back to India with Robert Plant.
Robert and I were just really keen to see what it would be like to go into a studio with some musicians from Bombay and see what we would come up with. We tried recording versions of "Friends" and "Four Sticks" with some percussionists, a half dozen string players, and a thing called a Japan banjo, and boy, was that tricky! They were great musicians, but they were used to counting and feeling rhythms in a different way. Moving them from one time signature and retaining the feel that we had in mind was difficult, but it was exciting and a learning experience. We tried "Friends" and "Four Sticks" because I thought it was safer to do something that Robert and I knew. It was easier to keep the thing rooted.

Rhythmic nuances between cultures are the hardest thing to teach.
But that's exactly what makes it interesting—it's the fusion. Remember what you were saying about the aggressive nature of the beginning of "Friends"? I had to get it right, dead on the beat, and they're doubling up and going all over the place. I'm thinking, "Oh, my God!" Still, it was really cool stuff.

Not everything's going to work. In order to create, you have to be able to fail sometimes.

Yes, but then you have to go away, you have to digest all the elements and think about how it could work in a particular song or a different set of circumstances.

We were seriously considering playing and recording in Cairo and India after those sessions. Peter Grant was definitely looking into it. We were investigating moving all our equipment via the Indian air force, but we were a little ahead of our time.

As you mentioned earlier, one of the songs you did work on out in Bron-Yr-Aur was "That's the Way," which is in a G tuning.

There's a number of tunings on the third album, but I was experimenting with altered tunings going back to the first album. The C tuning on "Friends" was something new, but the open G tuning [guitar strings, low to high: D G D G B D] on "That's the Way" was pretty conventional—Muddy Waters, Robert Johnson, and many others have used it.

"That's the Way," however, was very exciting to record because it gave us a chance to work with some new acoustic textures. John Paul Jones plays the mandolin on it, and the main breaks on it are taken up with the pedal steel. I couldn't really play a pedal steel like a pedal-steel player, but I could play it like me. And right at the end, where everything opens up, I played a dulcimer.

I never noticed the dulcimer on that—it blends so well with the guitar parts.

I was doing a bunch of overdubs and got excited. John Paul Jones went home, so I put the bass part on it as well! [*laughs*] That didn't happen often, believe me! The open tuning gave the track a lot of space, so we had a great time filling it up. And Robert's lyrics were superb.

Where did you record the album?

It was recorded in England. We primarily moved around between Island and Olympic, and the mastering was done at Ardent studios in Memphis with Terry Manning.

After the third album, you began using a mobile unit so you could record at locations in the countryside like Headley Grange and Stargroves. Did you start feeling that going into a traditional studio was killing the vibe?

My mind came to that conclusion, yeah. I didn't know exactly how the Band recorded their *Music from Big Pink* album or *The Basement Tapes*, but the rumor was they were done in a house they had rented. [*Music from Big Pink* was actually recorded in a traditional studio environment; *The Basement Tapes*, however, was recorded on a two-track machine at a house outside of Woodstock, New York, that Bob Dylan and the Band inhabited.] I didn't know for sure if they had, but I liked the idea. I thought it was definitely worth a shot to actually go someplace and really live it, rather than visiting a studio and going home. I wanted to see what would happen if all we did was have this one thing in sight—making the music and just really living the experience of it. I felt it would be important, and the hunch was right. The work ethic was pretty addictive. We knew what we were doing was right and that it was actually breaking new ground. We were cutting with a machete knife through the jungle, and discovered a temple of the ages.

Speaking of a "temple of the ages," around that time you bought a house in Scotland once owned by ceremonial magician Aleister Crowley. Additionally, the first pressings of the third album included the core tenet of Crowley's philosophy, "Do what thou wilt" and "So mote it be," inscribed on the lacquer during the final mastering process. Were your occult studies contributing to the vibe of your musical vision?

You could say that the inscription was a little milestone on the way—a point of reference. I sort of wondered how long it was going to take before anybody noticed. It took a long while. [*laughs*] No wonder they started playing our records backwards after that!

What do you think of the album artwork on *Led Zeppelin III*?

A disappointment. I'll take responsibility for that one. I knew the artist and described what we wanted with this wheel that made things appear and change. But he got very personal with this artwork and disappeared off with it. We kept saying, "Can we take a look at it? Can we see where it's going?" Finally, the album was actually finished and we still didn't have the art. It got to the point where I had to say, "Look, I've got to have this thing." I wasn't happy with the final result. I thought it looked teenybopperish. But we were on top of a deadline, so of course there was no way to make any radical changes to it. There are some silly bits—little chunks of corn and nonsense like that.

But it is no worse than my first meeting with an artist from Hipgnosis, who were the people that designed Pink Floyd's covers. We had commissioned them to design *Houses of the Holy*, and this guy Storm Thorgerson came in carrying this picture of an electric green tennis court with a tennis racket on it. I said, "What the hell does that have to do with anything?" And he said, "Racket—don't you get it?" I said, "Are you trying to imply that our music is a racket? Get out!" We never saw him again. We ended up dealing with one of the other artists. That was a total insult—*racket*. He had some balls! Imagine. On a first meeting with a client!

Were there any album covers that sparked your imagination when you were growing up?
I really loved this one Howlin' Wolf album that had a rocking chair and guitar on the cover. I don't know why it was so powerful for me, because it really wasn't such an amazing image. Maybe I just liked the music inside, and that made me like the cover.

There was also a John Lee Hooker album on the Crown label that had this great painting of a guitar on the cover that I liked. But, again, maybe it was just the music inside—it was definitely one of Hooker's best recordings. Usually in those days, I would've preferred to see a picture of the artist. With that in mind, it's odd that we rarely put our pictures on our covers. [*laughs*]

I want to talk about some of the instruments and the amps used during this time. It's interesting that you've created all this classic work using a relatively modest Harmony Sovereign acoustic guitar. It's not some extraordinarily rare thing like people would imagine, but a modest working tool.

Well, that's the guitar that I had. Martins weren't readily available in England back then. Gibson acoustics were starting to appear, but they were quite a lot of money, and I was quite happy playing my Harmony. "Babe I'm Gonna Leave You," "Ramble On," "Friends," and even "Stairway To Heaven" were arranged on that guitar. I didn't get my Martin until after the fourth album was released. I used the Harmony all the way through, really.

Where did you get it?
I don't remember now, but if I had to hazard a guess I'd say I probably bought it at one of the local music shops like Selmer's around the time of the Yardbirds, but I might have bought it a bit before then. Did I do a session on it? I don't know. The majority of my acoustic work in the studio was done on my '37 Cromwell f-hole guitar.

Mickie Most had kindly lent me his Gibson J-200 for the first album. That was a magnificent-sounding guitar, absolutely incredible.

It's funny, you still use the Danelectro that you've had since you were a session musician in the mid-sixties, and you've had your number-one Les Paul since 1969. You're extremely loyal to your guitars.
Yeah, that's quite right. If I were going out on tour tomorrow I'd be using the Les Paul I bought from Joe Walsh in 1969. I probably would still use the double-neck that I bought while in Zeppelin to play "Stairway to Heaven," as well. There's some sort of allegiance to those guitars—they're old friends.

I think that's great.
Well, traveling with them is not such a good idea, though. I was really attached to my black Les Paul Custom and I took it on the road with me in 1970 and it was stolen. I was using my Joe Walsh guitar, but I sort of plucked up the courage to take the Custom guitar on tour with us because when we did the Royal Albert Hall show, I'd used it to do those Eddie Cochran numbers and it was a real good backup guitar to have. It really sounded terrific. It was a leap of faith to take the guitar on the road, and look what happened. In spite of that, I've toured with the Joe Walsh

one ever since, as I say, so maybe that's something a bit peculiar in my makeup.

I think that people who are too fickle about their instruments are missing out. It takes time to understand what an instrument can do. I think there's something profound about exploring the capability of an amp or guitar and really understanding it, so you can do what you want with it.

It's true, if there's anything to try out, as far as a new amp or whatever, you can imagine what guitar comes straight out to try it. It's that number-one Les Paul, because I just know that guitar and it really knows me as well. It's fine. It works.

I wanted to ask you about one mystery regarding an amp you used in live performance around the time of the third album. Most people assume that you always played a Marshall live, but for a really crucial time—at the Albert Hall gig and at the Los Angeles Forum show in 1970, captured notoriously on the *Blueberry Hill* bootleg—you used a Hiwatt amp.

It was a transitional amplifier. I'm really reluctant to say what I used before the Hiwatt, because once I do, they'll get bought up and I'll never get my hands on one again! Oh, what the hell . . . I've got a couple of them. My main amp in the early days was a Vox Super Beatle, which was superb.

So, after the Vox, I looked around and everyone was using Marshall amps, so of course I wanted to do something different, so I got the Hiwatt, which had a foot-switch overdrive. Eventually I did go on to the Marshalls.

Interestingly, a little while ago before the Led Zeppelin reunion show, I set up all my amplifiers to see what was and wasn't working, and that Hiwatt kicked ass, man! It kicked ass.

MUSICAL INTERLUDE

A CONVERSATION WITH
JIMMY PAGE AND JACK WHITE

JIMMY PAGE AND JACK WHITE OF THE WHITE STRIPES
TALK ABOUT THE CONNECTION BETWEEN HIGH ART
AND THE LOW-DOWN BLUES.

————

AFTER GRADUATING high school in 1993, eighteen-year-old Jack White worked as an upholsterer in a gritty neighborhood of Detroit, Michigan. It was during this rather humble period, however, that White got serious about his music. He immersed himself in the primitive sounds of the Detroit garage-rock revival, playing drums and guitar in several bands while privately forming a very sophisticated musical aesthetic. The turning point came in 1996 when he married Meg White and began teaching her how to play drums.

Thus was born the White Stripes, a band whose music was informed in equal parts by the harsh sounds of thirties-era Mississippi Delta blues, Dutch minimalist art, the electric aggression of Led Zeppelin, and Meg's naïve drumming. The results were so original and uncompromising that soon the entire alternative-rock world stood up and took notice. At a time when the slick pop of Jennifer Lopez and Justin Timberlake ruled the airwaves, the White Stripes disrupted the charts with raucous slabs of artsy garage rock like "Fell in Love with a Girl" and "Seven Nation Army."

The White Stripes recently called it quits, but White soldiers on, performing as a solo artist, producing artists such as country legend Loretta Lynn, and running his own Third Man record label.

"I think it's magnificent how Jack has stood his ground," Page says of White. "Of course, you have to have talent. He is unique and has clarity of vision, which makes him refreshing. His honesty in his playing and approach is to be admired. Most musicians would've compromised. He hasn't and he won't. He's a solid rock."

Jack is equally glowing about the guitarist who he acknowledges had an enormous influence on his music.

"Jimmy Page has the special gift of taking an idea and presenting that idea in its most powerful form," White says. "Artists often lose their focus or become distracted, but that's never been the case with Jimmy. For example, as the Yardbirds were ending, he was able to find new people to work with, musicians that he knew could most powerfully present the ideas he had for the blues. What's even more impressive is that it was at a time when everyone thought that the blues had been taken to its highest, hardest-hitting point. It turned out to not be the case. Page came along with Led Zeppelin and turned it up ten more notches.

"I also believe that his work as a producer at times exceeds even his importance as a guitar player. Not only did he write incredible riffs, he also knew how to present them."

Jack, you've used primitive elements of the blues to rebel against what you perceive as an excessive and overly processed and technological culture.

JACK WHITE That's the main thing to rebel against right now—overproduction, too much technology, overthinking. It's a spoiled mentality; everything is too easy. If you want to record a song, you can buy Pro Tools and record four hundred guitar tracks. That leads to overthinking, which kills any spontaneity and the humanity of the performance.

What was interesting about Led Zeppelin was how well they were able to update and capture the essence of the scary part of the blues. A great Zeppelin track is every bit as intense and spontaneous as a Blind Willie Johnson recording.

Jack White and Page, 2006 (© Ross Halfin)

Led Zeppelin's version of Bukka White's "Shake 'Em on Down" on Led _Zeppelin III_ [entitled "Hats Off to (Roy) Harper"] is a great example of a track that captures the essence of the country blues without copying them.

JIMMY PAGE The key is you don't want to copy the blues; you want to capture the mood. On _III_, we knew we wanted to allude to the country blues but, in the tradition of the style, we felt it had to be spontaneous and immediate. I had this old Vox amp, and one day Robert plugged his mike into the amp's tremolo channel, and I started playing and he started singing. And what you hear on the album is essentially an edit of our first two takes. The band had an incredible empathy that allowed us to do things like that.

But that gets back to what you were saying before: You can't overthink this music. Mood and intensity can't be manufactured. The blues isn't about structure; it's what you bring to it. The spontaneity of capturing a specific moment is what drives it.

WHITE One thing is for sure: Jimmy doesn't get enough credit for his skill as a producer. Not only did he compose and play these great songs, but he was able to capture great performances from his band and made sure it was all properly recorded. I would go as far as to say that the way you were miking Bonham's high hat was just as important as how heavy your riffs were. You had an amazing sense of how to deliver that rhythm, not only in your guitar riffs but also in the production of the music. It was the culmination of all of these elements that made Zeppelin so dynamic.

PAGE I had this idea of making a collage of contrasting sounds to create a wide range of dynamics, right from the first album. It just evolved from there.

Jack, Jimmy went to art school and you've taken some cues from the world of fine art. Your second album was entitled *De Stijl* after the Dutch movement that attempted to purify art by bringing it back down to basic colors and form.

WHITE When we were finishing that album, I decided I wanted to dedicate it to Blind Willie McTell. [*De Stijl* features a cover of McTell's "Your Southern Can Is Mine."] During that time it hit me that McTell and most of the great country bluesmen were recording and performing in the early twenties, which was the same time period as when the De Stijl art movement was taking root. They were both doing the same things: breaking things down to their essences.

In my mind, both the country blues and the De Stijl movement represented a new beginning of music and art, perhaps for the rest of eternity. Both broke their respective arts down to its very core. You couldn't get any more simple and pure than the De Stijl school. They only used squares, circles, horizontal and vertical lines, and primary colors. That's it. The country blues of Son House and Charley Patton also brought music down to its fundamentals.

I wanted to draw those comparisons between those two things, which made people think that Meg and I were art students, which we weren't. I couldn't afford it. I probably would've gone if I could.

But you don't have to go to college to study or read about art.

PAGE I think that's a good point. If you are on to something creative, school can also inhibit you. The wrong teacher, man, can really mess you up.

How do you know if you are creating something important?

WHITE You know a songwriter's heart is pure when they have the desire to keep digging deeper into music. And invariably, when you dig deeper it always leads you into the past. Once I was able to dig back to the music of the twenties, it enabled me to understand more clearly the music of the present and the music that Jimmy was making in the Yardbirds and Led Zeppelin. It even helped me to understand my place in the musical universe. It's like we're all connected as a big gang of roving minstrels.

Jack, you said in a previous interview that it's easy to play like Stevie Ray Vaughan and difficult to play like Son House. Could you clarify what you meant by that?

WHITE I guess what I meant was, the blues scale is one of the easiest things you can learn on the guitar. It's the old cliché—"It's easy to learn but takes a lifetime to master." That's where I was headed with that. I'm not impressed with somebody playing a blues scale at blinding speed, but I am impressed with Son House when he plays the "wrong" note. Somehow it's more meaningful to me when I hear him miss a note and hit the neck of his guitar with his slide.

I think the distinction you're looking for is that Son House is not being superficial—he's not just playing a scale. He means every last note and is projecting it. He's not showing off his technique, he's trying to create a real emotional moment.

PAGE Technique plays a part—you have to know how to play. But what is important is that pursuit of something new and capturing that moment. Every band I've played in did a great deal of improvisation onstage, which is where the real magic takes place. That's where the real drama happens. You might fuck up, but that's also part of it. It's the tension that makes it exciting. Great music is never safe or predictable.

Jack, what impresses you about Jimmy's work?

WHITE I remember knowing the break in "Whole Lotta Love" when I was six. I had it on a cassette tape and there was actually a glitch on the tape from where the solo began because I had rewound it to that spot so many times. But now, as an adult, what impresses me is that Led Zeppelin is the ultimate expression of the power of the blues. Jimmy was really able to center in on the most powerful aspects of the form. If there was a knob for the power and expression of the blues, he was able to turn it up all the way.

I can give you an example of what I'm talking about. On the *Led Zeppelin* DVD, the band plays a version of "Dazed and Confused" on Copenhagen television that always gets to me. Right before the second verse, Jimmy starts making a bunch of abrasive noise for two seconds, and that is so much like a one-hundred-percent amped-up version of Robert Johnson. When Johnson did that sort of thing, it was the most powerful sound he could make using just an acoustic guitar and microphone, and when Jimmy did that, he was making the most powerful sound he could make in the environment he was in.

When you have a vision like Jimmy's, I think that's the aim. To make everything as powerful as you can make it.

PAGE But it wasn't just power—atmosphere was very important for us as well. We wanted to create an atmosphere that was so thick you could cut it with a knife. Our goal was to make music that was spine tingling.

Both of you are guitarists that produce. What does production mean to you?

WHITE What you said before: having a vision for what you want done. I was originally afraid that having too much control would come off looking too egotistical: "Songs written by Jack White, produced by Jack White, guitar played by Jack White," and so on. But what it really came down to was my belief that I knew what I wanted to happen and it was just more efficient to do it myself. I didn't want to discuss what I wanted to do with someone for an hour. I just wanted to do it.

PAGE That's it, isn't it? Who needs someone else getting in the way of the process? Even if you're wrong . . .

WHITE . . . at least it's my mistake.

PAGE Actually, I should say, even if you're not *right*. We must always remember, Jack, *the artist is never wrong!*

There is something else unusual that unites both of you: Zeppelin wrote lots of great riffs, great hooks, and refrains, but rarely did the band write what I would call a conventional chorus. Zeppelin's biggest hits—"Stairway to Heaven," "Kashmir," "Over the Hills and Far Away"—don't have choruses. The same is true of many of the White Stripes' biggest songs, including hits like "Seven Nation Army" and "Blue Orchid." Was that intentional?

PAGE Yes, it was done purposely. We wanted every part of the song to be important and have movement. There was no need to retreat to the security of having a big chorus in every song. If you emphasize one part of the song, it trivializes the rest of the music.

WHITE As far as I'm concerned, the riff in Led Zeppelin's "The Wanton Song," for example, *is* the chorus. It could go on for a half hour and I would be completely riveted and satisfied. It's so powerful and concise that it never gets boring.

PAGE A riff can take on the aspect of a chorus in a listener's psyche. When that happens, the whole song becomes one big chorus. The idea of a hypnotic riff as the prime mover of a piece of music has been around for a long time, whether you're talking about the Delta blues or music from Middle Eastern and African cultures.

Is there one piece of music that either of you can point to that in your mind represents some sort of Platonic ideal of a great song?

PAGE Picasso once said that a painting is never done. I sort of feel the same way about music. I would never say something is perfect. There are performances that can generate a lot of emotion in me when I hear them, but I can't say if anything is perfect.

WHITE That sounds good. I'm gonna go with Jimmy on that one!

[C H A P T E R]

THE COMPLETE STORY OF LED ZEPPELIN'S
MASTERPIECE, *IV*, THE RECORDING OF "STAIRWAY
TO HEAVEN," AND SURVIVING AN EARTHQUAKE.

Headley Grange, home of *Led Zeppelin IV*, with black dog (© *Ross Halfin*)

"THEY TOLD US WE WERE COMMITTING
PROFESSIONAL SUICIDE . . ."

I T WAS THE winter of 1971, and the executives at Atlantic Records were giddy with anticipation. At any moment, they were expecting Led Zeppelin to deliver a new album, and early reports were that it was their best work yet—a goddamn epic, it was rumored—and just in time for Christmas.

But the label's seasonal good cheer was quickly extinguished when the band's formidable manager, Peter Grant, made a rather frosty announcement. The band, he decreed, decided that its fourth album would have no title, no mention of the group on the outside jacket, no record-company logos or catalog numbers, and no musicians' credits.

Atlantic was flabbergasted. No title? No credits? All hell broke loose.

"They told us we were committing professional suicide, and threatened war," Page recalls. "But the cover wasn't meant to antagonize the record company—it was designed as a response to the music critics who maintained that the success of our first three albums was driven by hype and not talent. We wanted to demonstrate that it was the music that made Zeppelin popular; it had nothing to do with our name or image. So we stripped everything away, and let the music do the talking."

The music did much more than talk. It blew the roof off the dump. Released on November 8, 1971, the untitled album spawned a series of instant rock classics including "Black Dog," "Rock and Roll," and, of course, the mother of all FM rock ballads, "Stairway to Heaven." It is still a monster. As the third-largest-selling album of all time, the Zeppelin opus has sold more than twenty-three million units in the U.S. alone and still sells in the thousands every month.

When asked about the controversy surrounding the untitled album, now commonly referred to as *Led Zeppelin IV*, Page explains that "the band avoided overanalyzing their career. There was never any 'Let's do this' or 'Let's do that.' There was never any plan to conform to some idea. It was always about doing what came naturally at that point in time."

The saga of *Led Zeppelin IV* officially starts in December 1970 at Island Studios, located in West London. As the band members entered the recording facility during that bone-chilling month, their spirits were high, and with good reason: They had produced three consecutive platinum discs in as many years, and their concerts were breaking box-office records the world over. As Zeppelin's popularity soared, so did their ambition to top themselves. Unfortunately, the initial sessions at Island failed to yield any real mojo, so the members decided that it was time for a radical change.

"We thought it would be interesting to record someplace with some atmosphere and just stay there," Page says. "The idea was to create a comfortable working environment and see what would happen. Robert and I had written a lot of our previous album, *Led Zeppelin III*, in an isolated area in the Welsh mountains and really enjoyed the experience. It was very beautiful and there was nothing to distract us. This time we thought it would be fun to bring the whole band somewhere and hire a mobile unit to capture that moment in time."

Page had heard about an old house in the English countryside a few hours outside of London that Fleetwood Mac had used as a rehearsal space, and he decided to check it out. Built in 1795, Headley Grange was a rather large three-story stone structure that had originally been used as a workhouse for the poor and the insane. It was far from lavish, but its rough charm suited the guitarist just fine. Plus the two-hundred-year-old building offered something much better than creature comforts—it had *presence*.

"It was very Charles Dickens," Page says. "Dank and spooky. The room I chose to live in was at the very top of the building, and the sheets were always sort of wet. Headley Grange freaked Robert and John Bonham out, but I liked it, actually. I'm pretty sure it was haunted. I remember going up the main staircase on the way to my room one night and seeing a gray shape at the top. I double-checked to see if it was just a play of light, and it wasn't. So I turned around pretty fast because I didn't really want to have an encounter with something like that. But I wasn't surprised to find spirits there, because the place had a miserable past. One real positive outcome of us recording there is that I believe we revitalized the energy at Headley. The place became lighter as a result of our stay there."

But atmosphere wasn't the only reason Zeppelin decided to stay at the Grange. Page discovered that the place sounded great as well: "After I visited the house, I knew straightaway that the acoustics would be good."

Soon after, the guitarist booked the Rolling Stones' state-of-the-art sixteen-track mobile recording unit, hired ace engineer Andy Johns, and quickly went about the business of converting the rustic poorhouse into the world's largest recording booth. The band members set their instruments in the house's front room and stuck their amplifiers in various cupboards and stairwells to isolate them, and within days they were ready to make rock history.

"I think part of the sound of the album can be directly attributed to the fact that we were working in a fairly complex acoustic environment," explains Page. "The sound wasn't being created in a standard square box like you have in a traditional recording studio. We were continually moving amps and mikes around the house and creating new recording spaces, which I'm sure affects the listener on a subliminal, subconscious level. It's an idea that I had been developing since our first album, but working at Headley allowed us to take it to the next level."

While the setting at Headley Grange was conducive to any number of rock-star indulgences, the band was the essence of discipline. During

Zeppelin's month-long stay, they not only recorded much of the fourth album, they also polished off several songs that would later appear on their sixth album, *Physical Graffiti*, including "Down by the Seaside," "Boogie with Stu," and "Night Flight."

"What we found exciting about our time at Headley Grange was the ability to develop material and record it while the idea was still hot," Page says. "We were never a band that did ninety-six takes of the same thing. I had heard of groups that were into that kind of excess around that time. They had to work on the same track for three or four days and then work on it some more, but that's clearly not the way to record an album. If the track isn't happening and it starts creating a psychological barrier, even after an hour or two, then you should stop and do something else. Go out: Go to the pub or a restaurant or something. Or play another song."

The rough-and-tumble "Rock and Roll" was a prime example of this philosophy in action. As Page recalls, the band was working hard on one of the album's more complex tracks, "Four Sticks," and it soon became apparent that the session was going nowhere fast. To break the tension, John Bonham began pounding out the opening drum riff to Little Richard's rock classic "Keep A-Knockin.' " Spontaneously, the guitarist started playing a riff on his 1959 Les Paul that felt so good he had to say, "Stop, let's work on *this*." And by the end of the day, the band had given birth to what would become "Rock and Roll."

"That's how it was going back then," Page says. "If something felt right, we didn't question it. If something really magical is coming through, then you follow it. It was all part of the process. We had to explore, we had to delve. We tried to take advantage of everything that was being offered to us."

That included the piano-playing skills of the late Ian "Stu" Stewart, one of music's great, unsung talents. Stewart played piano with the Rolling Stones in their early days and eventually became Mick Jagger and Keith Richards's road manager. He also kept tabs on the Stones' mobile studio and made a point of dropping in on the Zeppelin sessions to see that everything was going well and that the recording unit was in working order. Never one to let good talent go to waste, Zeppelin immediately put Stewart to work.

"Stu was extremely shy, but he was an astonishing piano player," Page says. "He was a true boogie-woogie virtuoso. At Headley Grange there was this old piano that had fallen into a state of disrepair and was almost unplayable. It was so bad, we never even thought to use it. But Ian came in and just started improvising this amazing lick on it. So I went over to him and I did my best to tune my guitar to the piano, and the other guys started playing tambourine, handclaps, and stomping in the hallway, and before you knew it, we had recorded 'Boogie with Stu.' "

Later, when the band was putting the finishing touches on "Rock and Roll" at Olympic Studios in London, they enlisted Stewart once again. "John Paul Jones usually played keyboards for us, but he had no problem letting Stu play piano on 'Rock and Roll,' " Page says. "When you have the chance to use a master at a specific style, you just step back and enjoy what comes out."

But what became of "Four Sticks"? As the finished album bears out, Zeppelin had better luck at Island. Once again it was John Bonham who provided the creative spark that allowed the band to nail the devilishly elusive tune. It seems Bonzo had just seen his rival, drum virtuoso Ginger Baker, play the previous night with his band Air Force. The next day he came in the studio grumbling, "I'll show Ginger Baker something." Grasping four drumsticks—two in each hand—Bonzo started pounding out the signature groove with such propulsive force it knocked new wind into the song. Not to mention that his use of four sticks gave the tune its name.

The magic of Headley Grange, however, wasn't always in effect. While most of the songs on *Led Zeppelin IV* had their genesis in the country, many received their final polish in London. The rhythm track of "Black Dog," for example, was recorded at Headley, but Page's guitar tracks were completed at Island Studios.

Page recalls how the song came together: "John Paul Jones came in with the opening riff, which was quite good. I then suggested that we build a song similar in structure to Fleetwood Mac's 'Oh Well.' In other words, I wanted to create a call and response between Robert's vocal and the band. 'Black Dog' was a very tricky song to pull together, and if you listen closely to the album, you can actually hear Bonzo counting us in by clicking his sticks together right before each riff. It took us a fair amount of rehearsal to be able to attack the song properly."

While the band's performance at the Grange was powerful, Page was dissatisfied with the sound of his guitar. He considered using a "direct" recording technique that Neil Young had used to great success on his massive-sounding hit "Cinnamon Girl." The guitarist and engineer Andy Johns agreed that the experiment would be more easily accomplished at Island, and they postponed finishing "Black Dog" until they returned to London.

Once at Island Studios, they resumed work on what would become the album's first single. As Johns explained in a 1993 interview with *Guitar World*, the aggressive guitar snarl of "Black Dog" was created by plugging Page's sunburst Les Paul into a direct box, and from there into a mike channel on the studio's mixing board.

"Andy Johns used the mike amp of the mixing board to get distortion," Page says. "Then we put two 1176 Universal compressors in a series on that sound and distorted the guitars as much as we could and then compressed them. Each riff was triple tracked—one left, one right, and one right up the middle. The solos were recorded in a more standard way. I wanted something that would cut through the direct guitars—I wanted a totally different tone color. So I ran my guitar through a Leslie and miked that in the usual way."

The effect was stellar, but just a warm-up for what was to come. "Stairway to Heaven," like "Black Dog," was developed in large part during Zeppelin's stay at Headley Grange, but ultimately the band decided the complex song needed to be recorded in a proper studio. Page particularly felt it needed a controlled environment so that the band could perfect it.

"The rhythm track on 'Stairway to Heaven' consisted of me on my Harmony acoustic guitar, John Paul Jones on electric piano, and John Bonham on drums," Page says. "We really couldn't have done the acoustic guitar and drums at Headley; we needed a nice big studio. It's as simple as that.

"Although we recorded the song at Island, it was really created at Headley. I'd been fooling around with my acoustic guitar and came up with different sections, which I married together. But what I wanted was something that would have drums come in at the middle and then build to a huge crescendo. Also, I wanted it to speed up, which is something musicians aren't supposed to do.

"So I had all the structure of it, and I ran it by Jonesy so he could get the idea of it, and then on the following day we got into it with Bonham. My sharpest memory of working on 'Stairway' was Robert writing the lyrics while we were hammering away at the arrangement. It was really intense. And by the time we came up with the fanfare at the end and could play it all the way through, Robert had eighty percent of the lyrics done. It just goes to show you what inspired times they were. We were channeling a lot of energy."

"Stairway" is also the only song on *IV* on which you can hear all of Page's main guitars. He kicks off the immortal intro with his Harmony. The rhythm part was performed on a Fender electric twelve-string, and many of the ending riffs were played on Page's main ax, the '59 sunburst Les Paul given to him by Joe Walsh.

However, perhaps the most famous solo in rock history was improvised on the old Telecaster that Page used frequently on the band's first album: "I had the first phrase worked out, and a link phrase here and there, but on the whole that solo was improvised. I think I played it through a Marshall."

As for why the song has endured for close to four generations, Page ventures: "I think the lyrics are really great. It allows people to conjure up so many images in their minds. When one listens to records they always come up with their own concepts and visions, and 'Stairway' really allows for that. The fact that we printed [the lyrics] on the inner sleeve demonstrates what we thought of the song. But even with the lyrics printed on the sleeve of each album and CD, people still came up with their own interpretations. That's wonderful.

"I contributed to the lyrics on the first three albums, but I was always hoping that Robert would eventually take care of that aspect of the band. And by the fourth album he was coming up with fantastic stuff. I didn't really get involved, but I do remember him asking me about the 'bustle in your hedgerow' and saying, 'Well, that'll get people thinking.' But other than that . . . "

Finally, is there any truth to the rumor that George Harrison inspired the Zeppelin anthem? Page scratches his head and laughs in sudden recognition.

"You've got the right story, but the wrong song! George was talking to Bonzo one evening and said, 'The problem with you guys is that you never do ballads.' I said, 'I'll give him a ballad,' and I wrote 'Rain Song,' which appears on *Houses of the Holy*. In fact, you'll notice I even quote 'Something' in the song's first two chords."

There is little question that Jimmy Page is proud of "Stairway to Heaven," but his real enthusiasm is reserved for the album's grand finale, "When the Levee Breaks," an epic reimagining of the 1929 classic by pioneering female blues artist Memphis Minnie. To some degree, it's understandable. The song could be seen as the culmination of all of the ideas he'd been developing since the band's first album. Anyone who has spent any real time talking to the guitarist about recording will hear his mantra, "distance equals depth," at least once over the course of a conversation. And the drum sound on "Levee" is the definitive statement on Page's thesis.

CONVERSATION

Q:

HOW ON EARTH DID YOU
GET THE DRUM SOUND ON
"WHEN THE LEVEE BREAKS"?

We were working on another song in the front room of Headley Grange when a second drum kit showed up. Rather than stop what we were doing, we told the people bringing it in to just set it up in the entrance hallway. The hall was massive and in the middle of it was a staircase that went up three stories. Later, Bonzo went out to test the kit and the sound was huge because the area was so cavernous. So we said, "We're not going to take the drums out of here!"

Andy Johns hung a couple of M160 microphones down from the second floor, compressed them, added some echo and compressed that as well, and that was all we needed. The acoustics of the stairwell happened to be so balanced that we didn't even need to mike the kick drum. Jonesy and I came out in the hallway with our headphones and left the amps back in the room and banged out the rhythm track to "When the Levee Breaks" right then and there.

Did you stumble on the sound early, or toward the end of your stay at Headley?
I don't really remember, but my guess is that it was at the end. I think we already had a lot of work under our belts, because we did quite a few overdubs on that song, which is sort of unusual for the sessions at the house. We were focusing more on finishing the rhythm tracks, so we must've had most of those done.

Did you regret not stumbling over that sound sooner?
No. That would've been a trap. You wouldn't want everything to sound as big as that. It works because it's in contrast to everything else.

Haven't you always been a champion of ambient miking?
I knew the drum, being an acoustic instrument, had to breathe, so it was always paramount in importance that the studio had to sound good. The drums are the backbone of the band.

But while Headley was great for the drums, it wasn't always the best for the guitars. A lot of the guitars recorded there were used as guide tracks and re-recorded later at Island. However, "Levee" was an exception.

What tuning is that in?

That's my electric Fender twelve-string in open G. It sounds lower because we slowed the track down to make everything sound more intense. That's also part of what makes that track sound so huge. If you slow things down, it makes everything sound so much thicker. The only problem is, you have to be very tight with your playing because it magnifies any inconsistencies. It doesn't work the same way that it does when you speed something up. That makes everything sound tighter than it is. We used the same technique on "No Quarter."

"Levee" also features one of the album's more distinctive mixes.

The whole idea was to make "Levee" into a trance. If you notice, something new is added to every verse. Check it out—the phasing of the voice changes, lots of backward stuff is added, and at the end, everything starts moving around except for the vocal, which stays stationary.

THE LAST THING left to do on *Led Zeppelin IV* was mix it. Engineer Andy Johns convinced Page that the only place to go was Sunset Sound in Los Angeles, which was widely acknowledged as one of the world's cutting-edge recording facilities. Besides, Johns insisted, it's always "nice to go to L.A."

Unfortunately, things went badly right from the beginning. As soon as Page arrived, he was greeted with a good ol' California earthquake.

"I remember lying in bed while it was shaking up and down," Page says with a laugh. "I immediately flashed on 'Going to California,' where Robert sings, 'The mountains and canyons start to tremble and shake,' and all I could think was, Bloody hell, I'm not taking any chances—I'm going to mix that one *last*. Which I did!"

What happened next no one knows for sure, but Page has his theories. After he and Johns finished the mix, they brought the tape back to England. When the band listened back, everyone was aghast—the album was a muddy mess. It sounded almost as if the duo had simply forgotten to add any treble frequencies to the tracks. Accusations started to fly as the other three band members began to wonder out loud what exactly Johns and Page had been doing during their month's stay in sunny California.

"It didn't sound anything like it did in L.A.," recalls Page. "I was astonished. At the time there were all these stories of tapes that had been wiped by the magnets used on British subways. Johns said something must've happened to the tapes on the way back, because they've lost all of their high end. I still don't really know what happened. Maybe the monitors were giving us a totally false sound picture, because Sunset Sound had these real state-of-the-art monitors that were able to reproduce a big stretch of frequencies. Who knows?"

Page immediately sat down and remixed the whole album at Island. For the most part the sessions were a success, but for some reason the guitarist just couldn't recapture the mix they had in L.A. of "When the Levee Breaks." Instead of pulling his long black hair out, he decided to listen once again to the Sunset Sound mix to see if it might give him a clue as to what was going wrong. Magically, when he pulled out the Sunset tapes, he discovered to his amazement that the original "Levee" mix was completely intact and sounded absolutely amazing. So he kept it. It was the only mix from Sunset that made it to the album.

Over the last thirty years, perhaps no other album sleeve has been dissected more than the one found on the enigmatic untitled fourth Led Zeppelin album. However, only one man knows the real meaning of the hermit with the lantern or the old man with the bundle of sticks. Let's ask him.

CONVERSATION

Q:

MY INTERPRETATION OF THE *LED ZEPPELIN IV*
COVER IS THAT IT IS SIMPLY A REFLECTION
OF THE MUSIC INSIDE: A HARSH URBAN
LANDSCAPE JUXTAPOSED AGAINST
ANTIQUITY. IN OTHER WORDS, THE BIG-CITY
BLUES OF "BLACK DOG" VERSUS THE CELTIC
FOLK OF "THE BATTLE OF EVERMORE."

Hmm . . . I used to spend a lot of time going to junk shops looking for things other people might've missed. You know, I'd find all these great pieces of furniture, really fine Arts and Crafts things that people would just throw out. Robert was with me on a search one time, and we went to this place in Reading where things were just piled up on one another. Robert found the picture of the old man with the sticks and suggested that we work it in our cover somehow. So we decided to contrast the modern skyscraper on the back with the old man with the sticks—you see the destruction of the old and the new coming forward.

Our hearts were as much in the old ways as they were in tune with what was happening, though we weren't always in agreement with the new. But I think the important thing was, we were certainly keeping apace . . . if not going beyond it.

The typeface for the lyrics to "Stairway to Heaven" was my contribution. I found it in a really old Arts and Crafts magazine called *Studio*, which started in the late 1800's. I thought the lettering was so interesting I got someone to work up a whole alphabet.

What about the inside?
The inside cover was painted by a friend of mine, Barrington Colbys. It's basically an illustration of a seeker aspiring to the light of truth.

So which figure represents the band: the giver of light or the seeker of truth?
A bit of both, I suppose. A bit of both . . .

DURING THE SEVENTIES, LED ZEPPELIN BECAME THE
QUINTESSENTIAL ARENA-ROCK BAND. THEIR THREE-
HOUR DISPLAYS OF VIRTUOSO MUSICIANSHIP WRAPPED
IN GLAM AND EXPLOSIVE PYROTECHNICS WON THEM
THE RESPECT OF THE CRITICS. AND THEIR DEBAUCHED
BACKSTAGE EXPLOITS MADE THEM LEGENDS.

Page with violin bow, 1973 (© *Carl Dunn*)

"THE TOURS WERE EXERCISES
IN PURE HEDONISM . . ."

RIDING HIGH FROM the blockbuster success of 1971's *Led Zeppelin IV*, the band had become rock's biggest draw, eclipsing such established titans as the Rolling Stones and the Who. A May 5, 1973, show in Tampa, Florida, shattered what was then the single-concert attendance record set by the Beatles in 1965, and Zeppelin, the undisputed heavyweights of the music world, now flew from gig to gig on their own luxury jet with the band's logo emblazoned on its sides.

In March 1973, the band released *Houses of the Holy*, which is perhaps the most lighthearted album in the entire Zeppelin catalog. With the exception of the moody "No Quarter," the forty-minute recording is frothy, celebratory, and energetic.

"My main goal on *Houses of the Holy* was to just keep rolling," says Page, who once again produced. "Although everyone was clamoring for another *Led Zeppelin IV*, it's very dangerous to try and duplicate yourself. I won't name any names, but I'm sure you've heard bands that endlessly repeat themselves. After four or five albums, they just burn up. With us, you never knew what was coming. I think you can really hear the fun we had on *Houses* . . . and you can also hear the dedication and commitment."

Songs such as the reggae-influenced "D'yer Mak'er" and the off-kilter funk of "The Crunge" allowed Zeppelin's dry sense of humor to shine through their exotic stew of mysticism and American blues, representing something of a stylistic turning point for the band. In fact, the blues, which was a huge part of the band's early work, is all but absent on *Houses*.

But while *Houses of the Holy* is decidedly playful, it is also one of Zeppelin's most intricate albums, thanks to sprawling compositions such as "Over the Hills and Far Away," "The Rain Song," and the shimmering opening track, "The Song Remains the Same," which features some of Page's most dazzlingly fast guitar work. Even the seemingly straightforward "Dancing Days" features surprising layers of rich harmonies and some seriously tricky slide work if one takes the time to look beyond the happy-go-lucky lyrics.

The extroverted tone of *Houses of the Holy* crossed over to the band's revamped concert act. Where the quartet once wore casual hippie finery onstage, they now started to dress with a little more pizzazz for big arenas. Page in particular began wearing custom-made clothes that befit his image as an international guitar god, becoming a fashion trendsetter in the process. His matador-style jacket covered by brocaded hummingbirds particularly created a stir in fashion-conscious London.

The bigger venues also demanded a larger-than-life light show, and Zeppelin was determined to deliver. The 1973 tour now featured state-of-the-art effects such as dry ice, a hanging mirror ball, banks of colored lights, and strobes. Musically, the band presented every song as an event unto itself. "Dazed and Confused" was now twenty minutes of sheer evil, while the band performed (and fervent audiences received) "Stairway to Heaven" as a transcendent religious sacrament.

"In the early days of the band, we were quite contained," Page says. "But by 1973 we really had confidence. By then we were able not only to play but also project with it."

The band's exuberance extended offstage as well. Led Zeppelin created the template for rock and roll bacchanalia. For the next several years they wrote the book on wrecking hotel rooms and ravaging groupies, doing so, naturally, with their customary panache. Led Zeppelin's offstage exploits and excesses were the stuff of legend, particularly when they played Los Angeles and New York. How far did they go?

"As far as our imaginations would carry us," Page says. "Those really were the days of pure hedonism. L.A. in particular was like Sodom and Gomorrah, but it always had that vibe, even going back to the golden age of Hollywood in the twenties and thirties. You just ate it up and drank it down. And why not?"

But their decadence, prodigious as it was, would be of little interest if Led Zeppelin wasn't one of the greatest live rock acts ever.

Danny Goldberg, one of popular music's most prominent personal managers—his clients have included Nirvana, Sonic Youth, and the Allman Brothers, to name a few—worked as the band's publicist on that ground-breaking tour. What with their private jets, their semi-sordid rock lifestyle and sold-out shows, he remembers Zeppelin were magnets for press coverage. Ultimately, however, it was the music that sold the band.

"The main reason Zeppelin is great is because all four members of the band are incredibly talented," he said in a 2008 interview. "You could've built an entire band around any of the four of them. But the real key is Jimmy, who masterminded the band. He had so much confidence in his own abilities that he had no problem surrounding himself with these incredible players. Jimmy also had a vision of what was going on in rock and roll at the time and how to take it to the next level. After the sixties, Zeppelin was the beginning of the next chapter."

The 1973 tour's high point came in July with the group's final performances at New York City's Madison Square Garden, extensive excerpts of which appear in the classic concert film *The Song Remains the Same.* Page says that between shooting the film and the adrenaline of the shows, he didn't sleep for five days.

"We were in New York, we were making a movie and playing great shows, and it was difficult to shut down that kind of electricity," he said. "You'd try to go to bed, but most of the time you gave up, because it was more fun to go out and enjoy yourself.

"During a typical Zeppelin show there was such an intense exchange between the band and audience. The band set off the charge and the audience gave it back, and it just built through the night. That was the phenomena: that transmission."

In the following conversation, the evolution and trajectory of the band's legendary live act is traced via the footage collected on the *Led Zeppelin* DVD, the video compilation meticulously produced by Page in 2003. Clocking in at nearly five and a half hours, the two-disc set centers around four exceptional Zeppelin performances: their appearance at London's Royal Albert Hall in January 1970, just one year after the release of their debut album; footage from Zeppelin's July 1973 shows at New York's Madison Square Garden that was not included in the concert film *The Song Remains the Same*; their five-night run at London's Earl's Court in May 1975; and their attendance-shattering appearances at England's Knebworth Festival in August 1979, one year before Bonham's death. Each performance has been restored, remixed, and remastered in digital surround sound under Page's supervision. The DVD also includes some fascinating odds and ends taken from television appearances, promotional clips, and interviews.

CONVERSATION

Q:

THE *LED ZEPPELIN* DVD FEATURES VIRTUALLY
EVERYTHING THAT WAS EVER PROFESSIONALLY
FILMED OF THE BAND'S LIVE ACT. SO WHY
ARE THERE JUST FIVE HOURS OF FOOTAGE?
THAT'S A FAIRLY MODEST OUTPUT FOR A
GROUP AS SIGNIFICANT AS LED ZEPPELIN.

The answer to that is complicated. First, you have to transport yourself back to when we formed in 1968, and what was going on in England at the time. The BBC controlled radio programming in England, and progressive rock music was low on its list of priorities. As I mentioned earlier, whenever we performed, we always insisted on playing something short, like "Communication Breakdown," and something long, like "Dazed and Confused," so it would be a fair comment on where the band was at and where it was going. Obviously, the BBC wasn't going to let us do that on a Top 20 program. So our only option to promote ourselves on radio was to perform on a small assortment of live rock programs that allowed you to play two or three songs. We took advantage of those rare opportunities, but it still offered only a minimal amount of exposure.

Another way we promoted ourselves was on television, but the opportunities were practically nonexistent. The DVD lets you see a little of what we were up against. There's one clip of the band on Danish television that's pretty funny, because you can see that the audience is completely terrified of us. They're so intimidated that they don't know whether to watch or to run away.

We included another clip of the group playing a French show. They actually booked a Salvation Army band to play before us, so it's clear they didn't understand anything about us at all. Even though we did a sound check, the balance is terrible. So it came to a point where we decided that there was no point in promoting the band on TV. The way they presented us was crap, and the sound was an absolute disaster. In those days, TV viewers had the benefit of only a single two-inch speaker! And Zeppelin were not designed to come through a two-inch speaker.

Obviously, the band quickly figured out how to promote itself without radio and television.
That's right. We made our live shows the main thrust.

Was it the same in America?
Similar. The Yardbirds were probably even more popular in the U.S. than in Britain. We had toured there extensively, and that experience allowed me

to get in tune with the evolving tastes of the American market. I knew that there was this whole other underground scene happening that didn't care about hit singles.

Were the American FM radio stations supportive?
Yes, right from the beginning. By the time we got to the States, people were already becoming familiar with the first album, and that made it easier.

So your vision went beyond just the music. You saw on a sociological level the emerging importance of FM radio, album-oriented rock, and the growing concert industry. And you created a band that could take advantage of those changes.
That's right. For instance, we didn't release singles for the Top 20 stations in America, but we would release singles for FM album-oriented radio. We didn't have to create songs for a pop singles market—that would've been the kiss of death for us. But once people heard a new Led Zeppelin song on FM radio, they knew there was a new album. That's the way it translated.

The other thing is, by not officially releasing a traditional single, we forced people to buy albums, which is what we wanted. We wanted them to see the band's complete vision. Plus, we didn't have to conform to anybody's notion of what or how long a song should be. I mean, we weren't about cranking out three-minute singles. Even "Whole Lotta Love" didn't conform to that. It was a different thing altogether.

So if you were happy promoting the band through radio and live shows, why did you film your 1970 gig at the Royal Albert Hall?
We didn't really instigate it. It was early 1970, *Led Zeppelin II* had just come out, and we were really getting big. So this director from *Top of the Pops* [a British music-variety TV program] asked to record an entire show for a possible television special. He shot it, edited it, and presented us with a little four-song show reel a few months later. We watched the edit and it was good. But we were moving so fast at the time that when we saw it, the show already felt dated to us. It seemed so passé! So the project was shelved.

Did you forget about it through the years?

Not really. I had a copy of it at home, and a number of bootleg versions kept surfacing. Finally we had a band meeting to discuss doing something with the footage. Everybody liked the idea, so we bought the film, but then a new problem reared its head—we weren't sure where the eight-track recording was. It had been moved around so much through the years that it took a while to locate it. I finally found it at this massive film-storage location, which made me start thinking of other projects we had done over the years. I decided to make time to hunt them down and create an inventory.

But it still wasn't very easy to locate everything within this huge archive, because our stuff wasn't all in one place—they were located under different names and accounts. But there was this kid there that was fantastic and really helpful. He was a big fan of the band and had made mental notes about where Zeppelin-related things were located. So bit by bit we managed to locate all the stuff that was there.

We found the multitracks for all the band albums, we found all of the reels of the 1979 Knebworth shows and the 1975 Earl's Court shows, and all the additional stuff shot at Madison Square Garden that was edited out of our movie, *The Song Remains the Same*. And that was about everything.

In retrospect, considering how popular Zeppelin was, it's astonishing that this is all we have. But, as I was saying earlier, we were really picky about how our music and the band was presented. It was all a reaction to wanting our music to be offered in a proper fashion.

What do you remember about the 1970 Royal Albert Hall show?

It was a very prestigious place to play. I remember really working hard to fill up as much space as we could with the three instrumentalists.

You seem quite introspective during the performance. You're a different person from the showman that would emerge in 1973.

Albert Hall was a massive gig for us, and we really wanted to do the best we could. It was a magic venue. It was built in Victorian times and you go in there thinking about all the musical history that has preceded you. On top of that, it was something of a homecoming for John Paul Jones and I because we had both grown up around there. So we were all really paying attention to what we were doing. I agree with you—you can see the concentration.

Most bands hate having to play their biggest songs live because they get sick of them. Zeppelin, however, wisely used their best-known songs—like "Whole Lotta Love"—as launching pads for improvisation, and in doing so pleased their audience and kept the song interesting for the band.

"Whole Lotta Love" is the only song that is repeated on the DVD—it's there from the Albert Hall show and again from 1973. We had to play it because it was a signature song. It was even more of a signature song in England because they used it as the theme on *Top of the Pops*. The way you hear it at Albert Hall is pretty much how we played it back then. But by the time you hear it again in 1973, we've added all sorts of things to it.

"Whole Lotta Love" isn't the only thing that evolved. By 1973, the band started looking much more glamorous. You look a little shaggy in 1970.

Shaggy? If you look at pictures of me in the Yardbirds, I'm quite dressy. Hmm . . . I think there was a period there where I couldn't afford any clothes! I'll admit I wasn't the figure of sartorial elegance at Albert Hall! I'll admit that!

That's quite a sweater vest . . .

Actually, I've still got that! I could give it to the Rock and Roll Hall of Fame and they could put it on one of their funny little mannequins!

I don't mean to be obsessive about this, but . . .

Why not? I am!

By 1970, Led Zeppelin stopped using warm-up bands and began the tradition of being the only band on the bill.

Well, on the '68 and '69 tours we had a set comprised of songs from the first album, and it was a reasonably good set. But after we recorded the second album, we had a very, very good set, but it was quite long.

Remember, we weren't using the radio to promote our albums, so every time we released a new recording, we would insert some of those songs into our set. By the time we got to *Led Zeppelin III* we had more new material, plus our older material kept evolving, so we didn't want to throw anything out. We started at two and a half hours and grew to three, and some nights it went to four.

Did you ever regret starting that precedent?
Only if I wasn't feeling well. I mean, our shows were well paced. We were each given our moment in the set so the others could rest. I had my solo spot on "White Summer," and John had his keyboard bits, and Bonham had his drum solo.

The second DVD starts out with a quick montage of live performance clips from the '72 tour, and then officially kicks in with the 1973 outtakes from the *Song Remains the Same* movie. Why did you include the '72 footage?
There was this huge gap between the Albert Hall show and the movie, and during that time we had recorded two albums—*III* and *IV.* I wanted to show how dynamic that period was, so we laced some footage of the '72 tour over a live performance of "The Immigrant Song."

Even from the raw footage from '72, you can see we're becoming more animated onstage after being quite contained at Albert Hall. It's really very interesting.

The two 1972 Los Angeles concerts [featured in edited form on the triple-CD *How the West Was Won*] have been legendary among bootleggers for years. Why do you think those shows were so good?
Well, we always seemed to play better when we were in the States. We were cocky and we'd show off, and it was fantastic. We didn't have to worry whether our families were in the right seats or that our friends were in the right spot.

The L.A. show on the twenty-fourth at the Forum was almost four hours long, and the Long Beach show on the twenty-seventh was three hours long. I think the Long Beach show was shorter because we wanted to cut out earlier so we could go and hit the L.A. clubs. I'm not joking—that was the exact reason!

Led Zeppelin were the kings of L.A. during that period. Your excesses are the stuff of legend.
Yes, but sometimes I think our behavior was worse in Japan. We did things there that you just wouldn't believe. For example, there was a night when one of us got our clothes tossed out the window and that person took advantage of that opportunity to run around on the rooftops of Japan naked. Then there was a public phone that disappeared off the streets and was found outside our doorway with all sorts of money in it. Not to mention another evening when the beautiful hand-painted screens in our rooms were chopped up with a samurai sword. I mean, I'm giving you three examples, but all of those things happened in a forty-eight-hour period.

Night after night after night we had all of this stuff going on, and we got away with murder. In retrospect, our Japanese hosts were probably completely horrified, but they were so polite, they just kept bowing to us!

Where did you get the stamina?
I think there was an enormous amount of adrenaline that we were building up onstage, and we were just taking it offstage into the land of *mondo bizarro*. You know, you'd have someone riding a motorcycle through a hotel hall, but that would only be exciting for fifteen minutes, then it would be next and next and next.

How important were drugs to you during that time?
I can't speak for the others, but for me drugs were an integral part of the whole thing, right from the beginning, right to the end. And part of the condition of drug taking is that you start thinking you're invincible. I'll tell you something that is absolutely crazy. I remember one night climbing out of a nine-story window in New York and sitting on one of those air-conditioning units, and just looking out over the city. I was just out on my own and I thought it might be an interesting thing to do. It was totally reckless behavior.

I mean it's great that I'm still here to have a laugh about it, but it was totally irresponsible. I could've died and left a lot of people I love. I've seen so many casualties.

But when you're performing at a standard that is so incredible and intense, it can't help but change your psyche. Everything is changing and mutating every night. And sometimes you're having nights that you just can't believe—you just can't believe it. And that is why we started making recordings of everything, because it was just getting so interesting. It's going to affect you somewhat. We were abnormal people to begin with. [*laughs*]

Can you offer some insight into your growing interest in the occult during this period?
It's unfortunate that my studies of mysticism and Eastern and Western traditions of magick and tantra have all come under the umbrella of Crowley. Yeah, sure, I read a lot of Crowley and I was fascinated by his techniques and ideas. But I was reading across the board.

Remember, just about any British band that came out of the sixties worth its salt had a least one member that went to art college, which was a very important part of the overall equation. It wasn't unusual at that time to be interested in comparative religions and magick. And that's it. It was quite a major part of my formative experience as much as anything else.

Crowley, for the most part, is misunderstood. His message was of personal liberation. He encouraged people to ask what they really want out of life and encouraged them to do it. For example, he wrote about the equality of the sexes, and that was shocking in the Edwardian age. He wasn't necessarily waving a banner, but he knew women's liberation was inevitable. He was a visionary and he did not present his views gently. I do not agree with everything he said or wrote, but I find a lot of it relevant.

But it's not surprising that people link you to Crowley. It's known that you have quite a collection of Crowley artifacts.
Certainly. I made references to it in my music. I've always made clear references to the sources of my ideas. For example, in my fantasy sequence in *The*

Song Remains the Same, I made a very clear statement as to what was going on in my life by using tarot-card references and being a seeker of the truth.

Your rock and roll lifestyle actually allowed you to pursue the Crowley maxim of "Do what thou wilt" in a way that the average person could never imagine.
But what I was doing, I guess, was promoting *my* lifestyle. I wasn't really preaching, because it wasn't really necessary. My lifestyle was just my life-style. I didn't feel the need to convert anyone; it was just the way that my life was taking me at the time. At the end of the day, from this vantage point, it can either be glorified or criticized.

Did your lifestyle ever catch up to you?
By the time we hit New York in 1973 for the filming of *The Song Remains the Same,* I didn't sleep for five days! Everything was so exciting—why would you want to go to sleep? You might miss something. Plus it was the end of the tour and we were going home after those shows, so I don't think I wanted to give up!

I couldn't even think of that now, but that's how our life was. In fact, during that and subsequent tours, we became so full of adrenaline that we started sleeping less and less, and we had to resort to drinking and sleeping pills to calm ourselves down long enough to sleep.

But when I was off the road, the pendulum would swing the other way. I equally enjoyed our breaks. But then again, maybe I was recharging my batteries for the next time out! [*laughs*] I almost had this split personality. I really enjoyed having a stable home life.

What would you do?
I would create a balance by applying myself to writing and developing ideas for the next album. For example, all of the guitar parts and layers to "Ten Years Gone" were all worked out as demos at home. That kept me sane. There was this balance of going on the road and coming home to rest. But the thing is, my whole life was Led Zeppelin, and that's all there was to it—on the road or off.

On your 1973 tour you started using your own private plane, the *Starship*. Was that a good thing, or did it just guarantee that the party could continue and you'd never have a moment of rest?

No, it was a good thing. It was a place where you could bring your music and books and create some semblance of continuity as you traveled from city to city. However, [former tour manager] Richard Cole ran into one of the air hostesses on the *Starship* recently and she told him, "You know I made a lot of money off of you guys," and Cole asked her how. "Well," she explained, "when people on the plane used to sniff cocaine, they'd roll up hundred-dollar bills to use as straws. Then after they were high or passed out, they'd forget about the money. So we would go around and grab all the money that was laying around." That might've been true, but I'll tell you one thing: They never got any of my money! [*laughs*]

When people listen to Zeppelin these days, I don't think they realize how many new trails the band was blazing, not only musically but also socially. Long hair was still frowned upon, let alone sex, drugs, and loud music. Were you ever hassled?

We had a lot of problems in the early days. Our manager, Peter Grant, told me this one story about touring with the Animals in the south of the U.S. in 1965. The guy that drove their bus was black, and he swam with the band in a hotel pool. And because he was black, the hotel management drained the pool and scrubbed it. That wasn't so long ago. That happened in the sixties.

It was my dream to play Memphis. I grew up loving the music that came out of Memphis and Nashville. But it turned out to be really depressing. We arrived in Memphis and were given the keys to the city. It was funny, because the keys at that time were these little plastic things, not like the big honorary things you receive these days. The reason we were getting the keys was because the mayor was astonished at how quickly "this Led Zeppelin fellow" had sold out the local arena. It occurred to him, whoever this "guy" was, he must be important and that he should get the key to the city.

We got the keys in the afternoon, but I guess they didn't like the looks of us. Shortly after, we were threatened and had to get the hell out of town as soon as we were done with the show. I was really mad because there were all these places I wanted to go—Sun Studios, where Elvis had recorded, and so on. They didn't like the long hair at all, man. It was seriously redneck back then.

At the Memphis show, the kids were standing in their seats as they watched us, and they were getting beaten back down. We were part of a sub-culture they didn't want the kids to know about—hippies with long hair.

Then we played Nashville the following night. We were in the dressing room getting ready to go out and do an encore, and this guy walked in and unbuttoned his coat and said to us, "If you guys go back out there, I'm gonna bust your heads." And he wasn't kidding. They were really angry. I mean, what are you gonna do? You aren't going to go against an armed police force that wants to bust your ass. It was time to leave, and we didn't play in the South for a long time after that.

I won't even go into any more stories, but there are plenty. It was just so depressing, because I spent so much of my youth studying the music that came out of that region.

When you think about it, it's incredible how brave Elvis Presley must have been to challenge those social and sexual barriers in such a conservative area.
Think about it! Elvis started in 1954—that was more than ten years before we arrived. It's miraculous that he made it through! He had the hand of God over him. He really did. He was the one that brought it all together. He brought blues and race music to the white culture.

You met Elvis. What do you remember about your encounter?
We were invited to see him play and then invited back to a party afterward. We went up to his suite and his girlfriend Ginger was there with just a few other people. I can tell you, we were really nervous. When he came in the door, he started doing his famous twitch. You know, he didn't put that

on—that was something he really did! You might think that's funny, but it was real cool to us. It was a little awkward at first, because his music meant so much to us. But then somebody said, "You know that hot rod you drove in the movie *Loving You?*" And that was that—everybody just dove into the conversation, relaxed, and had fun. He was wonderful—a fantastic man.

While some of your lengthier songs ran over twenty minutes, your approach to jamming was quite different from what bands like the Dead or Cream were doing at the time. It wasn't just a matter of trading solos—your lengthy improvisations were almost like connected set pieces.

They were more like "unset" pieces. "Dazed and Confused" is the most obvious example. It had several different triggers throughout the song that the band used to signal the next section. If we hit one of the triggers, you would know that you had to start shifting gears. But between each cue, anything could happen.

There were some gray areas that—

Well, I wouldn't call them "gray." And they certainly weren't black-and-white. Actually, what we were doing was quite colorful! We wanted to express ourselves as musicians but we didn't want to bore people. We wanted to keep them on the edge of their seats.

Did you ever get lost?

Inevitably, when you work in uncharted areas, you can't always find your way. There might be some really odd moments, but those odd moments, if you work them the right way, can become quite interesting. Something unintentional will come out quite clever. That was the focus. I mean, it's astonishing, the ESP between us. There'd be cues, sure, but there were so many times we'd just hit together spontaneously. It was such a joy.

You were certainly respected as a musician, but you were also a showman. You wore your guitar slung low and set a trend that still exists today. In a way, the

only acceptable way for a rocker to wear a Les Paul is at the hips. Was it about image, or was it more comfortable for you to play it low?

You have to understand, while I was in Led Zeppelin I was living and breathing it. It became a lifestyle, including wearing my guitar low, gunslinger style. Rock musicians at the time were like outlaws, and as in the movies, our holsters just got lower and lower.

Your various stage outfits have also achieved rare iconic status. Who designed, for example, the black dragon outfit worn on the '75 tour?

It was made by a woman from L.A. named Coco. I basically outlined what I wanted on it. For example, I asked her to personalize the pants with astrological symbols—Capricorn, Scorpio rising, Cancer.

I still have that suit, and the amazing thing about it is that it still looks brand-new, like it just came off the peg. I did a lot of roadwork in that thing and it's in wonderful condition. Most of my other clothes ended up in tatters, but the dragon suit still looks great.

I thought about what I wanted on my stage clothes carefully. After Coco made the dragon outfit, I had her make my white suit with the poppy on it. I would wear the black dragon one night, and the poppy suit the next. It became a ritual for me.

Did watching the performances on the DVD bring back specific memories?

I'd seen the performances in the past, but I just sort of rushed through them. For the DVD I spent a lot of time watching each show, and they really did bring back memories and in many ways made this project a real labor of love. The Earl's Court show, in particular, hit home. I'd just gone through a divorce, and that was the last show of the 1975 tour. I remember I decided to travel, because there was nothing really keeping me home. Shortly after that, Robert had his accident, and things were never quite the same. [Plant and his family were severely injured in a car crash on the Greek island of Rhodes on August 4, 1975.]

Between the Albert Hall show and Earl's Court, you certainly had enough material to issue two DVDs. Why did you decide to release all of this material at once?
The idea at first was to put out just the Albert Hall shows. But that didn't make sense to me. I knew we had precious little visual documentation of the band, so I thought, Why not just put it all out and have a nice big package? Technologically, the time is right to present the material because you have a crisp digital medium and 5.1 surround sound.

I think there was hope from management that we would be able to create a DVD from each performance. But it became apparent that, for example, the Earl's Court show would be quite tedious on its own because it consisted of nothing but close-ups and tight shots.

Why is that?
Our shows at Earl's Court were the first time, to my knowledge, that anyone used back projection in concert. It's pretty commonplace now, but back then it was a pretty revolutionary thing to do. We projected close-up shots of ourselves onto the screen behind the stage, and it allowed those in the back of the hall to see us. But those ended up being the only images recorded from the shows, and because of the nature of back projection, they're all close-up shots.

With that in mind, it just seemed more entertaining to present these performances as an unfolding story. So what you have on the DVD is a real journey. It begins when we rejected television appearances and continues through to our last performance at Knebworth.

What I like about Knebworth is that it brings everything full circle. What I loved about Zeppelin was that it was always in a process of change and evolution. Even our oldest songs would differ from night to night. You mentioned using "Whole Lotta Love" earlier as a springboard to new things. Well, even when we did it at Knebworth, I came up with a whole new middle section for it, just to show people that we were still thinking about what we were presenting them.

There may be precious little film of the band, but it's been rumored that Zeppelin made a number of professional board recordings through the years. The 1972 shows from L.A. are excellent, but can we expect more live audio?

You're right, we did record a lot of shows, but many of the board tapes were stolen from me years ago. They were sort of "relieved" from my house in the early eighties when I wasn't there. All that stuff, along with the recordings of our rehearsals, were stolen and have surfaced as bootlegs, and it's a drag.

I discovered the theft around the time that I was working on my solo album, *Outrider* [1988]. I remember looking around for some demos and sort of wondering where all my tapes were. There was so much going on around my house and in my life at that time, I just figured they'd turn up somewhere. Well, they did turn up—as bootlegs! Someone who was pretending to be a friend stole the tapes.

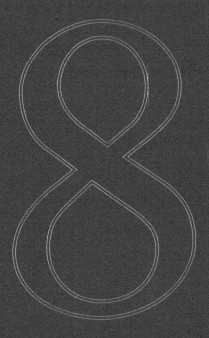

THE COMPLETE INSIDE STORY OF THE MAKING
OF THE *CITIZEN KANE* OF ROCK-CONCERT MOVIES,
THE SONG REMAINS THE SAME, INCLUDING THE
MEANING OF PAGE'S LEGENDARY FILMED FANTASY.

"IT WAS MY LIFE—THAT FUSION
OF MAGICK AND MUSIC . . ."

AFTER THE WORLD-BEATING 1973 tour ended in July, Led Zeppelin went uncharacteristically silent. It would be eighteen months before they would play another show, but during this lengthy sabbatical from the concert stage, Page was active on several fronts.

When he wasn't spending some much-needed quality time with his wife, Charlotte, and his three year-old daughter, Scarlet, Jimmy was working diligently in his home studio, demoing tracks for the next record. He also allowed himself some time to indulge in his growing passion for metaphysics by opening a small store devoted to occult literature. At the time, Page says, "There was not one place in London with a good collection of occult works, and I was tired of not being able to get the books I wanted." In the late autumn of 1973, he opened the Equinox bookshop at Number 4 Holland Street, just off Kensington High Street.

To help launch his new enterprise, the guitarist solicited the expertise of Eric Hill, a friend who worked at Weiser Antiquarian Books, a New York City bookstore world famous for its large antiquarian occult stock. Jimmy was a regular at Weiser's, and the two men had struck up an acquaintance over their mutual interests in books, art, and magick. Both were of the opinion that occult stores tended to be either, in Hill's words, "vulgar and sensational or dull and flat."

"We recognized the possibility of creating a shop with aesthetic refinement," Hill said in a 2008 issue of *Behutet*, a quarterly periodical devoted to modern magick and culture. "Our dreams at the time were fantastic."

It was a dream, however, that Page was determined to make a reality. After entertaining a number of locations, Jimmy found a suitably understated space near two beautiful parks and just down from Portobello Road, a street known for its antique dealers. In his quest for elegance and the appropriate ambience, he spared no expense in his store's layout, furnishings, and decoration.

"Jimmy hired a first-class architect to design the overall layout," Hill recalled. "The shop was partitioned with glass panels etched with Egyptian gods, including Thoth, Horus, as well as others. All the shelving and displays were rendered in a black neo-Egyptian art deco style, with soft lighting throughout. On the wall hung paintings by Crowley and Austin Osman Spare," an English artist, painter, and magician.

The name of the store was also thoughtful and evocative. Historically, the vernal and autumnal equinoxes are celebrated as times of spiritual balance and harmony—moments when night and day are of equal length. More to the point, *The Equinox* was also the title of a legendary series of Crowley-edited books, and it remains one of the definitive works on occultism and magick.

As defined by Crowley, magick was "the Science and Art of causing Change to occur in conformity with Will." Crowley saw magick as the essential method for a person to reach true understanding of the self and to act according to one's True Will, or grand destiny in life. This goal could be accomplished through forms of ritual, including invocation and evocation, astral travel, yoga, sex magick, and divination, among others.

In a 1975 interview with Page for the now-defunct *Crawdaddy* magazine, William Burroughs, the celebrated avant-garde writer best known for his harrowing novel *Naked Lunch*, explained it more succinctly: "The underlying assumption of magick is the assertion of *will* as the primary moving force in the universe—the deep conviction that nothing happens unless somebody or some being wills it to happen. To me this has always seemed self-evident. A chair does not move unless someone moves it. Neither does your physical body, which is composed of much the same material, move unless you will it to move. Walking across the room is a magical operation. From the viewpoint of magick, no death, no illness, no misfortune, accident, war, or riot is accidental. There are no accidents in the world of magick. And will is another word for animate energy."

Page's bookstore carried an eclectic array of esotericism, including volumes on Oriental philosophy, Kabala, tarot, alchemy, and Rosicrucianism. During the shop's lifetime—it closed in 1978—the Equinox also published new editions of two occult classics: the highly respected *Astrology: A Cosmic Science* by spiritual astrology pioneer Isabel Hickey, and Aleister Crowley and S. L. MacGregor Mathers's translation of *The Book of Goetia of Solomon the King.*

The astrology book was a worthwhile if somewhat conventional choice, but the idea of reproducing the *Goetia* was rather provocative. The word *goetia* is derived from a Greek word for sorcery, and it refers to a practice that includes the invocation of angels and the evocation of demons. *Ars Goetia* is the title of the first section of the seventeenth-century magick textbook *The Lesser Key of Solomon*; it contains descriptions of seventy-two demons that the biblical King Solomon is said to have summoned and then confined in a bronze vessel sealed by magick symbols. In 1904, Mathers and Crowley revised the *Ars Goetia* as *The Book of the Goetia of Solomon the King*, supplementing the text with Crowley's thoughts on ceremonial magick.

In addition to identifying a whole host of powerful and potentially destructive spirits, the book offered explicit instructions on how to conjure and banish them. In some ways, the *Goetia* was a dangerous and powerful little volume. Certainly Page knew this, and reproducing this important but combustible slice of arcana no doubt appealed to his outlaw sensibility and, perhaps, his sense of humor. But while reprinting the dark magick of the *Goetia* could have been construed as irresponsible, it would pose little threat to the general populace given the limited print run and the small audience of hipsters, scholars, and insiders who studied these things in the early seventies. (These days it can be downloaded from the Internet, free of charge.)

During this general period, Page also agreed to write the soundtrack for the film *Lucifer Rising* by underground movie director and noted occultist Kenneth Anger. By the time Page met Anger in 1972, the American-born director had built a reputation as one of the most influential independent filmmakers in cinema history. His groundbreaking surrealistic short films, like 1954's *Inauguration of the Pleasure Dome* and 1964's *Scorpio Rising*, had made him a legend among the art-house crowd and the toast of the emerging rock and roll counterculture. Like Page, Anger was also an admirer of Crowley and a follower of his religion, Thelema, so his work was often infused with sophisticated occult symbolism.

When Anger arrived in England to raise money for future productions, it was not terribly surprising that he and Page would cross paths. Page had seen several of his short films, including the Thelema-inspired *Invocation of My Demon Brother*, at a film society in Kent, and had avidly read an article in *Life* magazine from the mid-fifties that described Anger's visit to the small dwelling in Sicily that Crowley had used as an abbey in 1920.

"Crowley had been expelled from Sicily, and Mussolini's people had whitewashed the walls of the abbey, which Crowley had decorated for his better intentions," Page told journalist Peter Makowski in 2005. "When Anger visited the abbey, it belonged to two brothers: one was a fascist, the other a communist, and between them they had built a wall in the center of the room. The whitewash was still there, and there was this karma that continued on, as the brothers hated each other. Anger had gained access to the building to scrape the whitewash off the walls to reveal the murals and the frescoes.

"I could see that Anger was passionate about Crowley. So that, along with his creative output, made him somebody I would like to meet. Eventually, he came to my house in Sussex and I visited his flat in London."

During these initial social exchanges, the director outlined an idea for a film that became *Lucifer Rising* and asked Page whether he would like to create a soundtrack for it. The movie was to be a very condensed and coded work based on the Thelemic concept that mankind had entered a new period known as the "Aeon of Horus." It was a sort of New Age concept that celebrated individuality and "true will," as prophesized in Crowley's *The Book of Law*. Often misunderstood, the title figure of Lucifer is not evil or a devil but a bringer of light—a magical child who helps usher in the New Age.

The thought of collaborating with the respected artist on a subject that was near and dear to his heart was too much for Page to resist. He agreed to work with Anger and immediately started composing what has come to be thought of as one of rock's greatest lost treasures. At his home studio, Page began creating a lengthy, hypnotic, even trancelike, piece unlike anything he had done before.

"I employed a variety of instruments and effects," he explains. "I had a tamboura, which is an Indian instrument that produces a majestic drone. This was one I had brought back from my early travels in India, and it was about five and a half feet tall, and it was a really deep, resonant beast. It was the first thing I wanted to employ on [the track]. Then I had a Buddhist chant that was phased—nothing was quite what it appeared to be. I played some tabla drums, not very well, I might add, but the effect of it was really good. The one thing I wanted to do was avoid using guitar. There was a fraction of guitar right at the end, just a little taste."

In addition to working on the soundtrack, Page allowed Anger to use his state-of-the-art editing suite, located in the basement of his London home. Jimmy even made a brief cameo in the film, in which he holds a facsimile of the Stele of Revealing, an Egyptian tablet that was a central element in Crowley's religious philosophy.

Unfortunately, after Page had completed more than thirty minutes of music, he and Anger had an explosive falling-out. Accounts differ as to the cause of the split. Anger accused Page of taking too long to finish the soundtrack. Page, however, offered journalist Christopher Knowles a more detailed account.

"Kenneth had turned the basement of my home into an editing suite, and the housekeeper was there one day and found him giving some people a guided tour, and there was an argument. Kenneth took umbrage that he couldn't show people around, and the next thing I knew, I started getting all this hate mail directed at my partner and myself."

The tragic upshot was that Page's imaginative and genuinely haunting soundtrack laid dormant for almost forty years. When he finally released a limited vinyl edition of the *Lucifer Rising* soundtrack through his website, jimmypage.com, in 2012, it came as no shock that the music lived up to the composer's exotic description. Like smoke curling from a stick of burning incense, a multitude of textures slowly rise and unfold as tones produced by ancient and modern instruments; Indian ragas, Tibetan chants, and electronic effects intertwine until they overwhelm the senses. Like much of Crowley's writing, the music had one foot in antiquity and the other in some distant future, creating a truly disorienting effect on the listener. In many ways, *Lucifer Rising* is as important to understanding Page's aesthetic as any Zeppelin album. At no other time in his career had the guitarist let his imagination run this wild.

Despite these diversions, Zeppelin was never far from his mind, and Jimmy reconvened with John Paul Jones, Robert Plant, and John Bonham at Headley Grange in early 1974 to record basic tracks for what would become their sprawling, eclectic double album, *Physical Graffiti*.

Prior to *Graffiti*, producer Page had used the Rolling Stones Mobile Studio to record at Headley Grange, the band's informal studio and home away from home. Unfortunately, the Stones' unit was not available, so Page decided to use Ronnie Lane's mobile unit. Lane, bassist with the British band the Faces, built the studio with the aid of a young American engineer named Ron Nevison. Though Nevison was a little green, Page chose him to engineer the sessions for two reasons: He had just helmed the Who's intricate concept album *Quadrophenia*, and, as Page reasoned, "Who better to run a studio than the guy who built it?"

"The band was very prepared and well rehearsed," Nevison recalls. "There were very few takes, maybe six to eight, most of which were done to make minor adjustments to ensure the tempos were absolutely what they wanted and that the drum sound was correct. We'd start at one in the afternoon and end at around one in the morning.

"I started out living at Headley Grange with the band, but then I realized I would be better off not staying there. Sometimes it would get a little crazy. For example, they'd wake me up and want me to start recording something at four A.M. or something. I figured, for my sake and for theirs, if I wasn't around they wouldn't have that option. In the end, they agreed with me, because they knew they were a little insane."

Though much had been achieved during their stay, their sixth album was far from finished. Like *Led Zeppelin IV*, Page decided to record the final vocals and guitar overdubs in a more controlled studio environment. But before that could happen, other important business needed to be attended to—namely launching a new record company and finishing a documentary on the band that had been started months before.

In January 1974, Led Zeppelin's five-year contract with Atlantic Records ended, and manager Peter Grant renegotiated with the label for a handsome raise. Also included in the deal was an arrangement that all future Zeppelin recordings would come out on their own subsidiary label. In the spring, Grant and the band officially announced the formation of Swan Song Records, whose roster would include Zeppelin as well as veterans the Pretty Things, Scottish blues belter Maggie Bell, and a new supergroup, Bad Company, featuring Paul Rodgers of Free on vocals and Mick Ralphs of Mott the Hoople on guitar.

Page loved the idea of creating a safe haven for artists he respected, but he made it clear that he was more than happy to leave the business side of the label to Grant. It was a wise decision by a man who already had plenty on his plate.

Swan Song's official American launches took place in New York and later in Los Angeles in May 1974. By all accounts, both parties were lavish and decadent affairs with plenty of A-list guests, swan-shaped pastries, alcohol, and recreational drugs.

The launch of Swan Song was relatively smooth and hassle-free, but the same could not be said for the band's troubled documentary, *The Song Remains the Same*. The initial idea for the film was fairly straightforward—director Joe Massot and his crew were asked to shoot footage of the band's last three shows of the 1973 tour at New York's Madison Square Garden for a long-overdue Zeppelin concert movie.

"At the time we were interested in presenting the band on film," Page says. "We had already shot the Royal Albert Hall shows in 1970, but by 1973 we had moved on so far in such a short time that we felt the Albert Hall footage was passé in every respect. We looked and dressed differently, and the whole communicative quality of the music had been improved.

"We also felt we could do a more professional job, using multiple cameras and more sophisticated equipment. Prior to the Madison Square shows, the film crew came to two dates to prepare camera angles and gauge how much film they would need to shoot an entire concert. Unfortunately, after they finished shooting, we looked at the rushes and quickly realized that there were huge gaps in the filming. The crew hadn't covered basic things, like filming the verses to certain songs! We surmised that they were probably stoned—it was quite as simple as that. Everybody was stoned at the time, but at least we did *our* job."

It was at this time that the band came up with the idea for each member to film a "fantasy sequence" that would be used to cover the massive gaps in the film. Robert Plant would present himself as a mythic hippie hero rescuing a damsel in distress during "The Rain Song," John Paul Jones became a masked marauder whose exploits would be edited into "No Quarter," and to fill gaps in "Moby Dick," John Bonham would portray himself as a family man/farmer with a penchant for fast cars.

But of all the sequences, Page's was hands-down the most unconventional, reflecting his interest in tarot cards and the works of Aleister Crowley. Shot near his home and property in Loch Ness, Scotland—once owned by Crowley and purchased by Page in 1970—the scene depicts the guitarist climbing the steep, craggy mountain on a wintery, moonlit night. Decades earlier, Crowley himself would often make the same trek in the snow. Once he reaches the peak, Jimmy encounters an ancient hermit. When the hermit's face is exposed, it gradually becomes younger until it is clearly that of Page, but then further regresses to infancy until it reflects every stage in the life of man.

Despite director Massot's efforts to pull the film together, a rough cut of the movie was presented to the band and rejected. The director was fired and replaced by Australian filmmaker Peter Clifton, who convinced Zeppelin to reshoot their entire Madison Square Garden show on a soundstage at Shepperton Studios in England in order to fill the rest of the holes left in the original footage. To ensure some semblance of continuity, the band had to mime to the original Madison Square Garden recording, which proved to be difficult given the freewheeling and improvisational nature of their performance, but they gave it their best and the performances were remarkably convincing.

The Shepperton sessions, shot in the summer of 1974, were deemed a success, and the band instructed Clifton to stitch all the parts together and make a movie. Page knew that some things "would be completely out of sync," but he wasn't that concerned because "it was just something fun for the cinema." Besides, it was time for the band to move on, finish their new album, and plan another tour. The next time anyone in Zeppelin would seriously think about the movie again would be approximately two years later, in 1976.

Sessions continued for *Physical Graffiti* throughout the year, with additional overdubs and final mixing performed by Page and engineer Keith Harwood at Olympic Studios in London. Page came up with the album title, which represents the enormous creative and corporeal energy that had gone into producing and playing the set.

The goal of the new album, Page maintained, was to "keep that spark of spontaneity" at all times. "Tight but loose" was a phrase that Page often used to describe Zeppelin at their best, and *Physical Graffiti* was certainly that. Featuring some of the band's most ambitious and polished arrangements ("Kashmir," "Ten Years Gone," "In the Light") juxtaposed against its absolute grittiest ("In My Time of Dying," "Custard Pie"), the *Graffiti* was a sort of platonic ideal of his yin-yang vision.

"Throughout the album you could hear the four elements making up the fifth element," says Page of the sessions. "For example, 'In My Time of Dying' was recorded in one or two largely improvised takes. We're right on the edge. At over eleven minutes, it's the longest Zeppelin studio recording, but when you're playing like that, who wants to stop?"

Who wants to stop, indeed? The eight tracks recorded at Headley Grange and Olympic extended well beyond the length of a conventional album, so it was decided to include several unreleased songs from previous sessions and make the new album a double—a supercharged event that was, in Page's words, "guaranteed to knock your socks off."

Graffiti in many ways was a brilliant summation of where the band had been and where it was heading. It can be said that *Physical Graffiti* was the first album lengthy enough to display all of the facets of a band they had been developing since their debut in 1969.

The unusual tunings of the first and third album and epic reach of the fourth are heard on the otherworldly "Kashmir"; the driving, James Brown–influenced funk often evident in their live shows is unleashed on "Trampled Under Foot"; the pulverizing hard rock of the second album is echoed in "The Rover" and "The Wanton Song"; and the band's signature mysticism ("In the Light") and deep blues ("Custard Pie") permeate everything else. Even when *Graffiti* threatens to go over the top, its ambitions are pricked ever so delicately with a dash of quirky, light-hearted fun like "Boogie with Stu."

Page's guitar shines throughout, proving once again why he was universally hailed as rock music's best and most versatile ax man. "In My Time of Dying" celebrates his love of Delta blues with some truly hair-raising slide-guitar playing. His folk influences appear once again in the gentle instrumental "Bron-Yr-Aur," and "Houses of the Holy" demonstrates he could still compose direct, hooky riffs and play flat-out rock and roll with the best of them.

"I look at *Physical Graffiti* as a document of the band in a working environment," says Page. "Some people have said that parts of it are sloppy, but I think this album is really honest. It is a more personal album, and I think it allowed the listener to enter our world. You know, 'Here's the door. I'm in.' "

Both the name of the album and its packaging were meant by Page to directly reflect his own view of the record. The music, the title affirmed, was actual *physical graffiti*, filled with the band's palpable blood, sweat, joy, and pain. And the innovative album jacket was designed to literally beckon the listener to enter. Through die-cut windows, one could catch a glimpse of the group participating in the world's most outlandish and decadent party with guests King Kong, Elizabeth Taylor, Flash Gordon, and the Virgin Mary in attendance. The cover was meant to be a bit of elaborate fun, but it was also meant to mirror the band's increasingly bizarre real life in the rock and roll fast lane—a life that was about to get even faster and more weird.

Released on February 24, 1975, *Graffiti* was an immediate commercial success, becoming the first album to go platinum on advance orders alone. Shortly after its release, all previous Led Zeppelin albums simultaneously reentered the Top 200 album chart, making them also the first band in history to have six albums on the chart at once.

"It just made sense for it to be a double, given what Led Zeppelin were and how we worked," says Page. "It seemed like a very good idea. There may have been double and even triple albums out by other bands at the time, but I really didn't care, because ours was going to be better than any of them."

To satisfy their primal restlessness, Zeppelin hit the road two months before the release of *Graffiti*. It had been over a year since they had performed together, which seemed like an eternity. The tenth U.S. visit would feature a massive light show with a neon Led Zeppelin backdrop, a state-of-the-art laser show for Page's violin-bow segment in "Dazed and Confused," and seventy thousand watts of amplification that would guarantee crystal-clear sound even in the cheap seats.

In addition, Page premiered several new spangled outfits including his now iconic dragon suit, which he wore toward the end of the U.S. tour in L.A., and at the band's Earl's Court shows in the UK.

Their set was also revised to make way for all of the powerhouse numbers from *Graffiti*, such as "Trampled Underfoot," "Sick Again," and "Kashmir." Guitar hounds were particularly delighted when Page resurrected his Danelectro for the slide-guitar showcase "In My Time of Dying." But the band wasn't about to eliminate favorites like "Dazed and Confused," "Moby Dick," and "No Quarter," and their concerts regularly lasted well over three hours.

The early days of 1975 had gone wonderfully for Led Zeppelin, and they wanted to end the first half of the year on a high note. They decided there would be no better place to celebrate their continued success than at home with a five-night stay at the eighteen-thousand-seat-capacity Earl's Court Arena in London.

The band's entire U.S. stage was airlifted, at great cost, to England. Additionally, a huge Ediphor video screen, which would allow everyone in attendance a clear view of the performers, was erected above the stage for a cool £10,000. It was the first time the technology had been seen in England.

John Bonham later claimed that the mid-May London shows were among the group's best performances on UK soil, and many Zep connoisseurs concur. In addition to playing songs that were honed to perfection from its lengthy U.S. tour, the band rewarded its British audiences with something special: an acoustic set that had been dropped from the band's regular set since 1972. Page dusted off his Martin D-28 to play beguiling versions of "Tangerine," "Going to California," "That's the Way," and "Bron-Y-Aur Stomp," each adding a surprising touch of intimacy to the arena-sized event.

The sold-out concerts in their own country provided the band with a deep sense of pride in its accomplishments while offering a satisfying climax to what had been perhaps its best tour yet. Unfortunately, the good feelings would soon turn into a memory that would help Led Zeppelin weather the bad times that lay just around the corner.

A S LED ZEPPELIN prepared for an August tour of U.S. arenas, Peter Grant received some bad news. If the band wanted to hold on to the money they made over the previous year, they would have to flee England to avoid the country's punishing income-tax laws. While Bonham and Jones were not particularly pleased to have their family lives disrupted yet again, Page and Plant embraced the opportunity to travel and explore.

Page was recently divorced, so there was nothing to keep him home. Over the next couple of months the guitarist traveled to Switzerland and Rio de Janeiro, and by June he joined up with Robert Plant and his family in Morocco. They attended a music festival in Marrakech and visited the deserts of North Africa. Eventually, Page drifted off on his own to Sicily to investigate a villa that was up for sale near Cefalu, where Aleister Crowley had once lived in the twenties.

Plant, on the other hand, went to the Greek island of Rhodes. It was there that disaster struck. The singer, his wife, their two children, and Page's daughter, Scarlet, were involved in a serious car accident. Plant suffered multiple fractures of the ankle and elbow and his wife, Maureen, fractured her skull and broke her pelvis and leg. The children were also pretty banged up, and after many harrowing hours, the five were put on a plane and flown back to London for emergency medical care.

Led Zeppelin's summer tour was immediately canceled, and it wasn't certain whether Plant would ever walk again. But after Plant had time to recover, it was decided that the best therapy would be the one thing that always got the band through good and bad times: making music. In September, Plant, who was confined to a wheelchair, flew to California to meet up with Page in a rented beach house in Malibu.

For the next month, the two musicians worked together much as they had at Bron-Yr-Aur, coming up with song ideas that would eventually become their next album, *Presence*. In October, Bonham and Jones were summoned to the West Coast, where the band spent time at SIR Studios in Hollywood rehearsing the new material.

Four weeks later, the band flew to Musicland Studios in Munich, Germany, to record its seventh album. Page has often called *Presence* one of his favorites, both for its raw energy and because it represented an amazing triumph over adversity. In addition to working with Plant in his weakened condition, the band was also working under an almost impossible deadline.

"We only had eighteen days to record the whole thing because the Rolling Stones had time booked after us," Page says. "After we finished recording all of our parts, the engineer, Keith Harwood, and me just started mixing until we would fall asleep. Then whoever would wake up first would call the other and we'd continue to work until we passed out again."

Of course, Page could've continued somewhere else, but he felt the urgency helped him to create an interesting album. It was a true reflection of the level of their emotions. For the first time there were no acoustic songs, no keyboards, and no mellowness.

But there were lots of electric guitars.

Perhaps more than any album since their first, *Presence* is dominated by Page's searing guitar. Sensing that the band was in psychological disarray due to homesickness and the uncertainty of Plant's health, Jimmy rallied with an album's worth of unforgettable riffs. Some were sprawling and others were calculated to deliver a swift kick to the solar plexus, but all were designed to inspire the flagging group to pull up their boots and play as though their lives depended on it.

The jolting opener, "Achilles Last Stand," served notice that this was a band whose career was far from over. Clocking in at ten minutes and twenty-five seconds, this contender for Led Zeppelin's greatest studio moment features one of drummer Bonham's most powerful drum performances, a relentless, galloping bass line, and guitar parts so majestic and pit-marked that they successfully evoke the Greek ruins of Ephesus. Stir in some of Plant's best lyrics—a thinly veiled reference to his injury and mortality—and you can see why Page would call the song "very, very intense."

"I thought the solo on 'Achilles' was especially good," says Page, who is usually reluctant to pick favorites. "It's really singing out. When I listen back to it, I think to myself, 'My God, that solo says a hell of a lot to me. What was going on there?'" Page also gives a nod to the solo on "Tea for One," a wrenching minor-key blues on the album that reflected the isolation that every band member acutely felt due to their status as exiles.

Other notable songs on the album include "For Your Life," a venomous indictment of cocaine use in L.A., featuring a brilliant toothache of a riff played on a 1962 Lake Placid Blue Fender Stratocaster; and the exquisitely heavy "Nobody's Fault but Mine," an atomic rethink of the Blind Willie Johnson spiritual.

But not every song was filled with doom and gloom. The tongue-in-cheek "Royal Orleans" is an enjoyably loopy shaggy-dog story involving transvestites, while "Candy Store Rock" is a truly stellar tribute to all the early rocka-billy songs that Page and Plant loved as boys.

Still, there is no getting around it: *Presence* is the band's darkest and most personal recording. Plant called it "a cry from the depths," and Page admitted that "it's not an easy album to listen to." With its stark production, edgy guitar sound, and bleak worldview, it is no surprise that it was one of Led Zeppelin's least commercially successful releases. However, with each passing year, Page's insistence that *Presence* is among the band's very best work gains more credibility. Raw and real, the album is simply devastating.

Presence was officially released on March 31, 1976. Despite its jagged corners, the album still easily topped the charts in both the U.S. and the UK. Usually the band would immediately tour to promote its new music, but with Plant still on the mend, Page turned his attention to some unfinished business, namely the concert film *The Song Remains the Same* and its double live soundtrack album.

If Zeppelin couldn't go to the people, then the film would allow people to go to Zeppelin. Page knew that the movie had its flaws, but it also had considerable strengths.

Premiering in New York on October 20, 1976, and in London two weeks later—both times to lukewarm reviews—the film went on to become a major box-office hit, especially at midnight movie festivals, grossing an estimated $10 million by 1977.

CONVERSATION

Q:

FROM WHAT I UNDERSTAND, THE NOTORIOUS
FANTASY SEQUENCES FROM *THE SONG
REMAINS THE SAME* WERE CREATED TO
COVER GAPS IN THE PERFORMANCES DUE TO
MISCALCULATIONS MADE BY THE FILMMAKER.

Yes. It was our solution to that problem. The director, Joe Massot, was asked to work with the members of the band to develop their own segments.

Which was your favorite?
I really liked John Bonham's. It really captured his essence as a family man. It was fun, and the flipside of this roaring stage persona. In many ways, it reflected the way we all were at home.

How were the fantasy sequences developed? Did you guys discuss them with each other beforehand?
Not really. I knew what I wanted to do and Robert did, too—storming the castle and all of that.

When you saw the segments put together, did any of them surprise you? Was the band mutually respectful of one another's sequence?
In those days, I think being mutually respectful still meant there could be some piss taking. [*laughs*] I'm sure there were some nudges behind people's backs, and fair enough! I mean, it was hard to find the dividing line between doing a fantasy sequence in a rock and roll film and trying to be a *star* of the silver screen.

John's segment might've been fun, but yours is the most striking.
I had very strong ideas about my segment. I wanted to be filmed climbing this mountain face by my house in Loch Ness on the night of a full moon. Massot was astonished, because the night was perfect and the location was just how I wanted it to be. We shot it in December, so there was snow on the ground and these great clouds were going past the full moon. We created this scaffold for filming the shot, and everything was perfect and ready to go, but I'd forgot the most obvious thing; that I was going to have to do multiple takes climbing up and down. I kept thinking, What have I done! It was bloody cold up there, too, I know that much!

At one point in your segment, you're dressed as a hermit and you rapidly age into an old man. How was that done?

The transformation was done with a life mask [a mask produced from a cast of the individual's face], which I still have. Using that as a foundation, they created several different faces that showed me as I might look at various ages of life. I don't know how many there were, but there were quite a few. Then they joined all those shots of the different faces together.

When the film came out, I took my daughter, who was then six years old, to see it. That probably wasn't a great idea, because the film was so long and she was so young. But at the point where my transformation scene came on the screen, the theater was quiet except for this little voice that cried out, "That's not my daddy!" [*laughs*]

Could we talk a little about the meaning behind your sequence?

To me, the significance is very clear, isn't it?

I find it interesting that you were choosing to represent yourself as a hermit at a time when you were really quite a public figure.

Well, I was hermetic. I was involved in the hermetic arts, but I wasn't a recluse. Or maybe I was.

The image of the hermit that we used for the [inside cover] artwork on *Led Zeppelin IV* and in the movie studio has its origins in the painting of Christ called *The Light of the World* by the pre-Raphaelite artist William Holman Hunt. The imagery was later transferred to the Waite tarot deck [the most popular tarot deck in the English-speaking world]. My segment was supposed to be the aspirant going to the beacon of truth, which is represented by the hermit and his journey towards it. What I was trying to say, through the transformation, was that enlightenment can be achieved at any point in time; it just depends on when you want to access it. In other words, you can always see the truth, but do you recognize it when you see it or do you have to reflect back on it later?

There was always a certain amount of speculation about your passion for meta-physics. It may have been subtle, but you weren't really hiding it.
I was living it. That's all there is to it. It was my life—that fusion of music and magick.

Your use of symbols was very advanced. The sigils [symbols of occult powers] on *Led Zeppelin IV* and the embroidery on your stage clothes from that period are good examples of how you left your mark on popular culture. It's something that major corporations are aggressively pursuing these days, using symbols as a form of branding.
You mean talismanic magick? Yes, I knew what I was doing. There's no point in saying more about it, because the more you discuss it, the more eccentric you appear to be. But the fact is—as far as I was concerned—it was working, so I used it. But it's really not different than people who wear ribbons around their wrists—it's all a talismanic approach to something. Well, let me amend that: It's not exactly the same thing, but it is in the same realm.

I'll leave this subject by saying the four musical elements of Led Zeppelin making a fifth is magick unto itself. That's the alchemical process.

After you finished the fantasy sequences, you changed directors.
Yes. After inspecting all the footage, we discovered that we were still lacking. So the decision was made to hire a new director, Peter Clifton, and go into a British facility called Shepperton Studios. We re-created the Madison Square Garden stage and shot the remaining bits that we didn't have while miming to the original recordings. It was a good idea, but the only problem for me was figuring out how to mime my own lengthy improvisations. It was pretty impossible to do with any degree of accuracy. I suppose I could've gone back and learned how to play each solo note for note, but some of the instrumental breaks were over ten minutes long and I wasn't about to do that!

What made you finally release the project?
We were inactive after Robert had his terrible car accident in Rhodes, Greece, so we put it out while he was recovering.

In the end, were you glad you made the film?

Oh, yeah. In fact, there was always a desire to do another film. We talked about it in 1977. That would've been an interesting tour to capture, because it was extremely visual, and we were playing a lot of new material, like "Kashmir," "Achilles Last Stand," and "Nobody's Fault but Mine." I think you would've seen the same leap in our style and music from '73 to '77 that you saw from '70 to '73.

Is it true that some of the motivation to make a film stemmed from the fact that it would allow you to have more control over the sound than if you'd done a television special?

The sound was a major element of the movie. We had mixed it in surround sound, which was pretty cutting-edge stuff back then. Theaters in those days used three speakers; the center speaker for dialog, and left- and right-side speakers, which were used for effects. For *The Song Remains the Same*, we mixed the sound for five speakers and provided two additional speakers in the rear of the theater.

We used the rear speakers to create some really strong effects. For example, we had John Bonham's drum solo come out right over your head; and when I played guitar with a violin bow, we had the sound travel around the auditorium. People had heard music travel back and forth in stereo, but this was radical stuff for its day. We felt we needed some of those sonic highlights, because it was a very long film. Doing these effects was part of the pacing.

Considering how far audio technology has come since then, were you excited with the opportunity to revisit the mixes on *The Song Remains the Same* and reissue the film in 5.1 surround sound to re-create that experience for home theaters?

Sure. We already started in that direction when we mixed the *Led Zeppelin* DVD in 2003. Dolby surround sound has made a huge difference in things.

Were you able to readdress the syncing problems associated with the original print?

Yes. The big problem was we had to make the soundtrack fit the visuals.

Apparently, film is subject to really strong copyright laws, and it's almost impossible to even change a frame. To make the visuals sync better to the music, we had [engineer] Kevin Shirley move the music around with Pro Tools software. He really did a fantastic job. It's much better now.

But as I mentioned earlier, in the original film I'm out of sync a lot because I was trying to mime to my own improvisations at Shepperton, but it didn't look so obvious because everyone else was out of sync, too. Since Kevin was able to really tighten the vocals and the drums, now I *really* look out of sync!

When the movie finally came out, it was a pretty big box-office hit.
It was really gratifying. This was in the days before VHS tapes or DVDs, so the only place you could see it was at the movie theaters. It had a big cult following, and people would see it multiple times at midnight movie festivals. It was like the *Rocky Horror Picture Show*.

It was probably hard to get Led Zeppelin concert tickets in the seventies. The movie was the only way many people could see the band.
That's why we did it. It made sense to do it. But as usual, whenever we worked with people outside of our core group, it was a shambles. We did our best to pull it together, and it required a lot of imagination to salvage what could've been a disaster.

It's always harder than it should be to get people to put the same care into a project as you would.
If you watch carefully, you'll see a great example of that sort of carelessness in the film. Before I went onstage, I warned all the cameramen to stay away from me within reason, because I didn't want to be distracted while I was trying to perform. Of course none of them listened, and at one point you see this guy with a camera coming up to me and he's stepping all over my wah-wah pedal! You can hear it going up and down, so I just carried on using that wah-wah sound. What else are you going to do? It's warts and all, the whole damned thing!

Watching the film, I was impressed by the amount of precision, finesse, and control you applied to working the volume and tone knobs on your guitar. It's almost a lost art.

First, you have to be lucky enough to have an amp that operates on the threshold of clean and dirty, so that it can interact with the controls of the guitar. Once you have that, then you can start really playing with the volume and control.

It's different these days because there are so many ways to create guitar sounds, but back in the seventies you had to use what little you had to the greatest effect. All I had to really work with was an overdrive pedal, a wah-wah, an Echoplex [tape delay], and what was on my guitar. It wasn't a lot, and I had to create the entire range of sounds found on the first five Zeppelin albums. With that in mind, the volume and tone controls, and how and where you picked, were quite important.

How did the rather lengthy live improvisations on songs like "Dazed and Confused" and "No Quarter" develop?

When you're playing with a band that was as good as we were, you didn't want to stop after a one-minute solo! And look, if you're playing the same songs night after night on a long tour, improvising was a way to keep the music alive and interesting for yourself. I never wanted the songs to settle in. I've always enjoyed living by my wits with regard to my guitar playing. That goes back to even my session-musician days, where I had to come up with parts on the spot.

People have complained to me through the years that I never played the solos from the albums live, particularly on something like "Stairway to Heaven." That's why I decided to play it note for note at the reunion show in 2008—just to prove I could do it!

What I like about improvising is that great music is about tension and release, and sometimes you pull something out and sometimes you don't. It's not exactly a failure when you don't play something great; it's more like a heroic glitch! Your chance of success is greater, though, when you're surrounded by other great musicians, like I was.

Did you prepare for the film? Were you concerned about playing your best for posterity?

No, it wasn't like that at all. I think the only way I prepared for the filming was by staying up for five days straight! That's the truth.

When you looked back and revisited the soundtrack and the movie, did something stand out for you?

I thought "The Rain Song" was really good. I bet you didn't expect me to say that, but it has a real drama to it. It's not as good as the studio version, but I think it has its own character. I also like the bowed section on "Dazed and Confused," which really went well with the fantasy sequence.

Who re-haired the violin bow that you destroyed night after night while playing "Dazed and Confused"? Fixing a bow is not something just any roadie can do.

As you know, new violin bows are expensive, so what we would do is buy a bunch of warped ones and take them on the road. They were much cheaper!

The '73 tour was in support of *Houses of the Holy*, which followed the rather monumental fourth album. Did you feel any pressure to live up to standards set by that album and "Stairway to Heaven"?

Of course, but we didn't let it get in the way. My main goal was to just keep rolling.

One of my favorite songs on *Houses of the Holy* is the epic "The Song Remains the Same."

It was originally going to be an instrumental—an overture that led into "The Rain Song." But I guess Robert had different ideas. You know, "This is pretty good. Better get some lyrics—quick!" [*laughs*]

I had all the beginning material together, and Robert suggested that we break down into half time in the middle. After we figured out that we were going to break it down, the song came together in a day. I used a Fender Electric XII twelve-string guitar on that track. Before that, I used a Vox twelve-string to record things like "Thank You" and "Living Loving Maid" on the second album.

Did you keep a notebook or tape ideas?

I always did that. And then I'd patch them together later. I always had a cassette recorder around. That's how both "The Song Remains the Same" and "Stairway" came together—from bits of taped ideas.

Houses **is such a bright-sounding album. Did you vari-speed the tape up a notch to get everything to sparkle a bit more?**

No. The only song I can think of that we vari-speeded up were a couple of overdubs on "Achilles Last Stand." However, I applied the vari-speed to the overall track of "No Quarter." I dropped the whole song a quarter tone because it made the track sound so much thicker and ominous.

The upbeat vibe of *Houses of the Holy*—aside from "No Quarter"—suggests that you were feeling very positive at the time you recorded it. "The Crunge," for example, is a complete goof.

I played a Stratocaster on that one—I wanted to get that tight James Brown feel. You have to listen closely, but you can hear me depressing the whammy bar at the end of each phrase. Bonzo started the groove, then Jonesy started playing that descending bass line and I just came in on the rhythm. You can really hear the fun we were having on *Houses* and *Physical Graffiti*. And you can also hear the dedication and commitment.

If *Houses of the Holy* was one of your tightest productions, then *Physical Graffiti* is one of your loosest. Did you make a conscious decision to retreat from a highly polished sound?

Yes, but not completely. "In My Time of Dying" is a good example of something more immediate. It was just being put together when we recorded it. It's jammed at the end and we don't even have a proper way to stop the thing. But I just thought it was so good. I liked it because we really sounded like a working group. We could've tightened it up, but I enjoyed its edge. On the other hand, "Kashmir," "In the Light," and "Ten Years Gone" are all very ambitious.

Did you ever a force a song, or did you discard ideas that didn't automatically click?
We forced things on occasion. Actually, "When the Levee Breaks" off the fourth album is a good example. We tried "Levee" in just an ordinary studio and it sounded really labored. But once we got Bonzo's kit set up in the hall in Headley Grange and heard the result, I said, "Hold on! Let's try this one again!" And it worked. But we were never a band to try ninety takes. If the vibe wasn't there, we tended to drop it.

You and Plant were traveling to places like Morocco and the Sahara Desert around this time, and I can really hear that influence in songs like "Kashmir." Whose idea was it to explore Morocco?
I did a joint interview with [Beat novelist] William Burroughs for the American music magazine *Crawdaddy* in the early seventies, and we had a lengthy conversation on the hypnotic power of rock and how it paralleled the music of Arabic cultures. This was an observation Burroughs had after hearing "Black Mountain Side," from our first album. He then encouraged me to go to Morocco and investigate the music firsthand, something Robert and I eventually did.

You've said in the past that *Presence* is one of your favorite Zeppelin albums.
I guess it's because we made it under impossible circumstances. Robert had a cast on his leg and no one knew whether he would walk again. It was hairy!

Triumph over adversity—
That's exactly it. It was a reflection of the height of our emotions of the time. There were no acoustic songs, no keyboards, no mellowness. We were also under incredible pressure to finish the record. We did the whole thing in eighteen days. I was working on average eighteen to twenty hours a day.

It was also grueling because nobody else really came up with song ideas. It was really up to me to come up with all the riffs, which is probably why *Presence* is so guitar heavy. But I don't blame anybody. We were all kind of down. We had just finished a tour, we were nonresident, and Robert was in a cast, so I think everybody was a little homesick. Our attitude was summed up in the lyrics on "Tea for One."

What's your strongest memory of that time?
Fighting the deadline. We only had three weeks to work because the Rolling Stones had time booked after us.

Didn't you have the power at that time to demand more time from the record company to finish the album?
Of course, but I didn't want to. I didn't want the record to drag on. Under the circumstances, I felt that if it had dragged on, a negative, destructive element might have entered the picture. The urgency helped us to create an interesting album.

John Paul Jones's contribution to *In Through the Out Door* seems to be more significant than on other albums. Did you feel that it might be more interesting for you to function as an accompanist rather than at center stage?
See, you had a situation with *Presence* where Jonesy didn't contribute much to the songwriting, and that was a strain. I mean, I would've preferred having some input at that point. But he bought a new synthesizer [a Yamaha GX-1] and it inspired him to come up with a bunch of things for *In Through the Out Door.* He also started working closely with Robert, which was something that hadn't happened before.

Were you losing your enthusiasm for the band?
Never. Never. In fact, Bonzo and I had already started discussing plans for a hard-driving rock album after that. We both felt *In Through the Out Door* was a little soft. I wasn't really very keen on "All My Love." I was a little worried about the chorus. I could just imagine people doing the wave and all of that.

And I thought, "That's not us." In its place it was fine, but I wouldn't have wanted to pursue that direction in the future.

Led Zeppelin accomplished so much. Didn't you ever want a hit single?
No, not really. We just wanted to write really good music that would hold up on its own. Chart music tends to be a little disposable.

How would you like people to see your role in Led Zeppelin?
Many people think of me as just a riff guitarist, but I think of myself in broader terms. As a musician, I think my greatest achievement has been to create unexpected melodies and harmonies within a rock and roll framework. And as a producer, I would like to be remembered as someone who was able to sustain a band of unquestionable individual talent and push it to the forefront during its working career. I think I really captured the best of our output, growth, change, and maturity on tape—the multifaceted gem that is Led Zeppelin.

MUSICAL INTERLUDE

A CONVERSATION WITH
LED ZEPPELIN PUBLICIST DANNY GOLDBERG

IN 1973, DANNY GOLDBERG WAS A YOUNG MAN PUT IN
CHARGE OF CHANGING LED ZEPPELIN'S PUBLIC IMAGE.
HIS GOAL? MAKE 'EM BIGGER THAN THE BEATLES.

———————

D ANNY GOLDBERG is currently one of rock's most respected
personal managers, but once upon a time, over the hills and far
away in 1973, he worked for Solters and Roskin, an old-school
show-biz public relations firm.

"I was the resident long-haired rock and roller," said Goldberg. "So
when Led Zeppelin came knocking, I was immediately elected to represent
the band."

Goldberg said his marching orders were clear. Zeppelin were selling
more records and more concert tickets than any other band. But the Rolling
Stones were getting all the attention, and Zep manager Peter Grant wanted
something done about it.

Until 1973, Led Zeppelin had all but turned their backs on the press.
DANNY GOLDBERG That's right. They were not critically embraced when
they first came out. The critics loved Eric Clapton and Jeff Beck, and they
considered Jimmy Page to be somewhat of an interloper. The press had hurt
the band's feelings. But Zeppelin became instant superstars in the United
States due to FM radio airplay and their incredible live shows, and they felt
they didn't need print recognition.

When I started working for them, they were working on their fifth album, *Houses of the Holy*, and they were in a different frame of mind. They wanted a fresh start and sensed it was a new chapter in the band's history and that it was time to reach a wider public.

What was your first press release about?
Mostly about the size of their audiences. They played an arena in Tampa, Florida, that was slightly bigger than Shea Stadium in New York, where the Beatles held a record for the most attendance for a single artist. So our first press release was about how Zeppelin had broken the Beatles' record. It was an angle that worked, and the press all over the world picked up that Zeppelin were bigger than the Beatles.

Later in the summer, the band started using its private airplane, the *Starship*, and that became our angle, because it was a novelty and a lot of journalists had never been on a private plane. It was a good story and it fed the narrative that Zeppelin was a really big band.

Was the decision to make *The Song Remains the Same* also part of the narrative?
It probably came from the same impulse—to leave a more vivid footprint.

What do you remember about the director, Joe Massot?
I just remember that the band was mad at him all the time. There was some big drama that he didn't get all of "Whole Lotta Love" on film. Those were the days when you had these huge canisters of film on your camera, and when you'd run out of film, it took time to reload. He didn't calculate things properly, and the film ran out during the song. But you have to give Joe some credit: He did capture those performances, and that movie is something the band is remembered for. For many fans, it's their main experience of Zeppelin.

At the time, the film didn't seem very meaningful. It actually seemed amateurish and weird. I would never have predicted that it would be a film that people would be looking at thirty-five years later. But I was wrong.

Led Zeppelin and their Starship, 1973 (© *Bob Gruen*)

You weren't a fan of the film?

I was disappointed. I had such a vivid memory of the real performances, and those were better than the movie. But in retrospect, I'm really glad it exists. It's just so precious to have some documentation of those shows. I've bought many copies of it through the years, and I have a fourteen-year-old son that loves the movie. Now I see the band's impulse to make it was absolutely right, but at the time I didn't think it was really worthy of them.

What do you think makes the band so great? Why do we still care?

The main reason is that all four members of the band are incredibly talented. John Paul Jones, for example, had the lowest profile, and he was an incredible genius. Jonesy was the least well known, and any band would've killed to have him. There's no question that John Bonham was the greatest rock drummer who ever lived. And Robert Plant turned out to be an amazing front man, lyricist, and singer. Jimmy was the man with the vision, but the rest were great, too.

They were harder to imitate than you would think. There are a lot of bands that try to replicate the heavy element of their music, but they miss the light and shade—the sexuality, the brutality, and the sensitivity. It turns out to be a very difficult thing to copy. There are only a handful of artists that are so unique that they endure for decades, and Led Zeppelin are on that short list.

Yet so many critics would have you believe that they were only party animals.
They had a clarity and balance that was pretty sophisticated. They were always very serious about their music. For all the partying, all the tragedy, all the drama and flamboyance, they agonized over every detail. They did meticulous sound checks; they rehearsed; they worried about the lights, the sound, the set list. There was nothing lax about the way they did things.

I remember times when Robert had a cold and couldn't hit some of his notes, and he would be depressed for days after. Bonham would do hours of sound checks to get his drums to sound right for a show. Their success was no accident. They didn't just get stoned and improvise. I mean, they did get stoned sometimes, but they were always well prepared when it came to their music, and they took it as seriously as any painter or artist.

MUSICAL INTERLUDE

TEN TOP LED ZEPPELIN GUITAR MOMENTS

GUITAR MUSICOLOGIST JIMMY BROWN HAS TRANSCRIBED
EVERY GUITAR NOTE IN LED ZEPPELIN'S CATALOG.
HERE, HE PICKS TEN SONGS THAT DEMONSTRATE
WHAT MAKES JIMMY PAGE'S PLAYING UNIQUE.

———

MUSIC EDITOR AND guitarist Jimmy Brown, a man with two of the finest ears in the world, was charged by Alfred Music—the print publisher of all Led Zeppelin songbooks—with the formidable task of transcribing and putting on paper every note Jimmy Page played on the band's studio releases. Drawing on more than three decades' worth of documentation, interviews, filmed footage, recordings, and personal analysis, Brown completed the Page-approved catalog in 2012.

As a respected music theoretician and someone who is intimately familiar with the nuances of Page's playing, Brown was eminently qualified to select ten of Jimmy's finest performances and get the essence of what makes them so special.

1. "Since I've Been Loving You" (*Led Zeppelin III*)

In his rhythm-guitar part to this Chicago-style slow blues, Page plays an inventively slick turnaround phrase at the end of each chorus (initially from 1:06–1:12) that mimics a country-style steel guitar, with a bent note woven into and placed on top of each chord. What makes this phrase so interesting and enigmatic is how, over a D *flat* major seven chord (D♭maj7, played on organ by John Paul Jones), Page bends a C note up to D *natural*—the flat nine of D♭maj7—and manages to make it sound "right." It's something few musicians outside Miles Davis would have the guts to do.

2. "Dazed and Confused" (live version, *The Song Remains the Same*)
This ambitious twenty-eight-minute performance marks the apex of this song's evolution and features some of Led Zeppelin's most intense, creative interplay. Page is at the height of his powers here in terms of both chops and musical vision. The otherworldly violin-bow interlude, beginning in earnest at 9:10 and spanning nearly seven minutes, is particularly inspired.

3. "Achilles Last Stand" (*Presence*)
This is Page's crowning achievement in guitar orchestration. The song really begins to blossom at 1:57, and from this point on Page spins numerous melodic variations over the underlying Em-Cadd9#11 chord progression. Thoughtful consideration was put into the stereo image of each guitar track, which keeps the entire recording crisp despite the dense arrangement. As a side note, Page previewed this song's jangly, plaintive Em-Cadd9#11 chord figure in the previously mentioned 1973 live version of "Dazed and Confused," beginning at 5:52.

4. "The Rain Song" (*Houses of the Holy*)
Played in an unusual tuning (guitar strings, low to high: D G C G C D) with lots of ringing open strings and unison-doubled notes, this song features a sophisticated chord progression that was inspired by Beatle George Harrison, who challenged Page to write a ballad. After playfully evoking the verse section of Harrison's "Something" on the first three chords of "The Rain Song," Page veers off into an ultimately more ambitious progression.

5. "Kashmir" (*Physical Graffiti*)
Played in D-A-D-G-A-D tuning, which Page had previously used to great effect on "Black Mountain Side" from Led Zeppelin's debut album, "Kashmir" is built around four mesmerizing riffs, three of which involve the use of

open-string unison- and octave-doubled notes, which create a natural chorusing effect and a huge, epic sound. Particularly noteworthy is the way Page overlaid, at 0:53, the song's ascending main riff—the James Bond–theme flavored part—with the recurring descending sus4 chord sequence. Page remarked, "The descending chord sequence was the first thing I had. After I came up with the 'da-da-da, da-da-da' part, I wondered whether the two parts could go on top of each other, and it worked! You do get some dissonance in there, but there's nothing wrong with that. At the time, I was very proud of that."

6. "Whole Lotta Love" (*Led Zeppelin II*)

One of the heaviest intro/verse riffs ever written! Not content with playing it "straight," as his blues-rock contemporaries might have, Page inserts a subtle, secret ingredient into this part: Instead of playing the riff's second and fourth note—D, on the A string's fifth fret—by itself, he *doubles* it with the open D string, then proceeds to bend the fretted D note approximately a quarter step sharp by pushing it sideways with his index finger. The harmonic turbulence created by the two pitches drifting slightly out of tune with one another is abrasive to the sensibilities and musically haunting, but the tension is short-lived and soon relieved, as Page quickly moves on to a rock-solid E5 power chord. "I used to do that sort of thing all over the place," he explained once. "I did it during the main riff to 'Four Sticks' [from *Led Zeppelin IV*], too." In that song, the guitarist bends the G note at the D string's fifth fret by pushing it with his middle finger while sounding the open G string.

7. "Going to California" (*Led Zeppelin IV*)

Page also used unison strings to great effect on this acoustic masterpiece. Playing with both his low E and high E strings down to D (in what is known as double-drop-D tuning), the guitarist plays dreamy hypnotic arpeggio figures that feature lots of ringing, repeated notes played on different strings. With its blend of English and American folk-guitar styles (think Bert Jansch meets Merle Travis), "Going to California" is a finger stylist's delight. Particularly compelling is the dramatic bridge section beginning at 1:41, played by Page in the parallel minor key, D minor. If you listen closely, you'll hear two acoustic guitars fingerpicking different inversions of the same chords, thirds apart.

8. "Babe I'm Gonna Leave You" (*Led Zeppelin*)

Another acoustic masterpiece, this song features a bittersweet circular chord progression presented as ringing, fingerpicked arpeggios. Particularly noteworthy is the way Page spins numerous subtle melodic variations on the theme throughout the song (check out the one at 3:40), sweetening the aural pot with dramatic dynamic contrasts. This may be one of the most perfectly recorded and mixed acoustic guitar tracks ever. Notice how, in the song's intro, the "dry" (up-front and un-effected) acoustic guitar is in the left channel while the right channel is mostly "wet," saturated in cavernous reverb.

9. "Stairway to Heaven" (*Led Zeppelin IV*)

Page ran roughshod all over two rules of pop music with this masterpiece: It's more than eight minutes long, a previously prohibitive length for pop-radio formats, and the tempo speeds up as the song unfolds. "Stairway" is the epitome of Page's brilliance not only as a guitarist but also as a composer and arranger, as he layers six-string acoustic and twelve-string electric guitars throughout a beautiful composition that gradually builds to a crescendo before culminating in what many consider to be the perfect guitar solo.

10. "Over the Hills and Far Away" (*Houses of the Holy*)

This song is a study in contrasts, specifically between English-Celtic-flavored acoustic folk and Gibson Les Paul–driven hard rock. It begins with a playful folk-dance-like riff, which Page initially plays on a six-string acoustic and then doubles on an acoustic twelve-string, that gives way (at 1:27) to crushing power chords and a clever single-note riff built around pulled string bends (first heard at 1:37). Particularly cool is the way Page reconciles this

218 LIGHT & SHADE

electric riff with the strummed acoustic chords introduced earlier. To top it all off, Page, the producer, concludes the song with a false ending; as the band fades out at 4:10, a lone guitar emerges with a final variation of the folk riff from the intro, but all you hear is the 100-percent-perfect "wet" reverb "return" signal, which creates a mystical, otherworldly effect.

[CHAPTER]

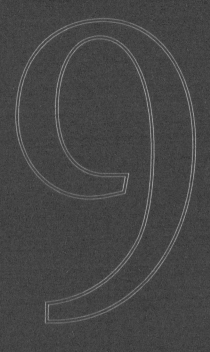

9

THE DEATH OF DRUMMER JOHN BONHAM LEADS TO
THE END OF LED ZEPPELIN. PAGE REGROUPS AND
MOVES "EVER ONWARD, EVER ONWARD."

Jimmy Page, 1984 (© Neal Preston)

"I WAS KNOCKED SIDEWAYS . . ."

WITH *IN THROUGH THE OUT DOOR* completed, manager Peter Grant wanted to let the music industry know that, despite their four-year hiatus from the British concert stage, Led Zeppelin was still the biggest band in the Western world. Grant and promoter Freddy Bannister worked together to have the group headline two massive shows at Knebworth Park, in Hertfordshire, on August 4 and 11, 1979. It was also decided that the album would be released to coincide with the shows to take advantage of the publicity surrounding the concerts.

More than one hundred thousand fans attended, and considering how long it had been since Led Zeppelin had rock and rolled, the Knebworth performances were surprisingly strong. "Achilles Last Stand," "In the Evening," and "Rock and Roll" still had plenty of the old magic, as can be seen on the official *Led Zeppelin* DVD. Plant looked a little older and Page appeared dangerously thin, but overall the band looked happy, and there was cause for optimism. Despite all they'd been through in recent years, all four members still wanted to play music together and carry on as a group.

After Knebworth, a small European tour was booked, and it was such a positive experience that the band agreed to once again plan an autumn tour of North America. But it was not to be. On September 24, 1980, drummer John Bonham attended a rehearsal at Bray Studios for the upcoming tour. After a full day of playing and drinking vodka, he returned to Page's new house in Windsor with the rest of the band. At midnight, one of Jimmy's assistants led Bonham to an upstairs bedroom, where the drummer went to sleep. Sometime within the next few hours, the drummer's heart stopped beating. John "Bonzo" Bonham, the very foundation of Led Zeppelin, was dead; he was thirty-two years old. The coroner ruled it "death by misadventure," concluding that Bonham had died from choking on his own vomit while asleep due to alcohol consumption.

A few weeks later, the devastated band members met with Grant at the Savoy Hotel in London, where Plant told their manager they simply couldn't go on without Bonham. It wasn't surprising. Page often described Led Zeppelin's mysterious chemistry as four elements coming together to create a powerful fifth. It stood to reason that if you eliminated one component from the equation, the whole structure would collapse. "If it had been any of us, I don't think we would have continued," Page says. "It wasn't that sort of band. Nobody else had the ability that John had."

On December 4, 1980, the surviving members issued a public statement that made it official: Led Zeppelin was no more.

Page says he was "knocked sideways" by Bonzo's death, calling it the worst time in his life. Speculation swirled regarding the guitarist's escalating drug use, and his increasingly gaunt appearance did little to make people think otherwise. It was evident that the hitherto unshakable rock star was shattered. But in time he started playing again and even making the odd guest appearance with other musicians. In March 1981, he joined Jeff Beck for a surprise encore at the Hammersmith Odeon. That he was determined to move on became absolutely clear when he acquired the Sol recording studio, located approximately ten minutes from his home, from Elton John producer Gus Dudgeon.

While there was little doubt that he would eventually resume his career as a creative musician, questions—When? How?—remained. Opportunity came literally knocking on his door in the form of his neighbor Michael

Winner, a film director best known for his hugely successful 1974 vigilante shoot-'em-up, *Death Wish*, starring Charles Bronson. Winner was in the process of making a sequel when he approached the guitarist and composer to write the soundtrack.

The project had to be completed in eight weeks, which was just the sort of extreme challenge Page needed to rouse him from his bereavement. Film scoring was a relatively new medium for Jimmy, but, undaunted, he went at it with his customary authority. Written and recorded in the fall of 1981 at the Sol, the musically diverse *Death Wish II* score was impressive by any standard. "Who's to Blame," sung by ex-Colosseum/Atomic Rooster vocalist Chris Farlowe, and "The Release," a taut, guitar-driven instrumental, sounded like logical extensions of *Presence* and *In Through the Out Door*. But it was "Prelude," a bluesy interpretation of Chopin's Prelude No. 4 in E Minor for piano, and the eerie "Hotel Rats and Photostats" that demonstrated that Page was once again his old self, unafraid of charting entirely new directions.

It was rumored that the *Death Wish II* music editors had some doubts about Page's technical ability to deliver the goods. As it happened, his work was so polished that Winner declared it to be the most professional score he'd ever worked with.

After Page completed the *Death Wish II* project, Atlantic Records insisted that Led Zeppelin was contractually obligated to deliver a final album. After briefly considering a live collection, the band decided to scour the cupboards and deliver a disc of outtakes and experiments instead. As producer, it was Page's job to assemble, organize, and edit the pieces—to turn them into a genuine album. After *Physical Graffiti*, little was left in the Zeppelin vaults, but Jimmy managed to find enough music to create a pleasingly scattershot collection of curios. The best of the lot were the seriously raucous "Wearing and Tearing," culled from 1978's *In Through the Out Door* sessions, and the audacious "Bonzo's Montreux," a 1976 Bonham drum instrumental electronically treated by Page.

Although Page was quite active musically, he worked mostly out of the public eye, and the world began to regard him as a bit of a recluse. The man who had been, for nearly a decade, the central—and very visible—leader of rock's biggest band appeared to many to have dropped from the face of the earth.

All of that, however, was about to change. In May 1983, Page joined Eric Clapton for encore performances of "Further On Up the Road" and "Cocaine" at the Civic Hall in Guildford. But what looked to be a spur-of-the-moment casual jam was in fact a carefully planned prelude to one of the most unexpected and thrilling rock and roll events of the eighties: a superstar charity concert tour featuring, for the first time together on any stage, the legendary guitar triumvirate of ex-Yardbirds Page, Clapton, and Jeff Beck.

The tour, which benefited Britain's Action Research into Multiple Sclerosis (ARMS), was the brainchild of Ronnie Lane, ex-bassist for the Faces and himself a victim of the disease. Things got underway with a pair of shows, presented on successive evenings, at London's Royal Albert Hall. Revenues from the first performance went to ARMS while those of the second, which was attended by Prince Charles and Princess Diana, were earmarked for the Prince's Trust, a charity providing support for disadvantaged young people.

Along with the three guitar legends, the shows featured a who's who of British rock musicians, including Traffic vocalist Steve Winwood and the Rolling Stones rhythm section of bassist Bill Wyman and Charlie Watts, among many other friends and colleagues of Lane.

As expected, both nights were special, with each player giving his all for Lane and others suffering from MS. After rousing sets by Clapton and Beck, Page, in his first major post-Zeppelin appearance, was greeted with a long and emotion-filled ovation. Clearly, the audience was thrilled to see Jimmy back in action. Backed by Simon Phillips on drums, Fernando Saunders on bass, Chris Stainton on keyboards, Andy Fairweather-Low on guitar, and Winwood on vocals, Page played three songs from the *Death Wish II* soundtrack—"Prelude," "Who's to Blame," and "City Sirens"—using a brown 1959 Telecaster that had made its first appearance on the 1977 Led Zeppelin tour. The short set concluded with a majestic instrumental version of "Stairway to Heaven," played by Jimmy on his iconic Gibson double-neck.

The crowd responded ecstatically, their cheers shaking stately Albert Hall to its most royal foundations.

"I was terrified," Page says of his comeback. "But I wanted to do all the shows. It was funny—I agreed to participate, but at the last moment I thought, 'What on earth am I going play?' Everyone else had such notable solo careers. In the end, all the musicians made it easy because they were working so tightly together for the cause."

The concerts were an unqualified success, and it was decided that the show would travel to America for dates in Dallas, San Francisco, Los Angeles, and New York. Unfortunately, Winwood was not available, so Jimmy asked Bad Company's Paul Rodgers to take his place on vocals.

For the U.S. shows, Page and Rodgers added two songs to the set: "Boogie Mama," a song from Rodgers's recent solo album, and a new joint composition entitled "Bird on a Wing" (later to be renamed "Midnight Moonlight"). Almost six years had gone by since Page had last toured North America, and, as was the case in England, he received a hero's welcome. Page later said that the shows did him "a world of good . . . the fans wanted me back."

The ARMS shows reignited Page's passion for the stage, and he was determined to find a new vehicle for his musical ambitions. His chemistry with Rodgers on the U.S. dates had been promising, and both men felt that a project was worth exploring. Bad Company's own days were numbered and, fortunately for Jimmy, Rodgers was open to new possibilities.

In the summer of 1984, the duo recruited a young fretless-bass player named Tony Franklin, who'd been playing with Jimmy's friend Roy Harper, and drummer Chris Slade, who had played with Pink Floyd's David Gilmour and Manfred Mann's Earth Band. They christened the new quartet the Firm, and Page says at the time that their goal was to do nothing more than have fun and "just get out and play."

By November 1984, the band had amassed enough original material to tour Europe, and three months later they released a self-titled album to coincide with a tour of North America. The idea of the principal figures of Led Zeppelin and Bad Company joining forces was simply too much for U.S.

audiences to resist, and the tour was a hit. The album and shows were further buoyed by their first hit single, "Radioactive," which mixed Paul Rodgers's trademark soulful vocals with an irresistibly dissonant, funky edge.

The Firm served as a way for Page to break with the past and embrace the future. This was made clear by the band's courageous decision not to play any Led Zeppelin or Bad Company material in concert. They weren't interested in earning easy applause and seemed genuinely determined to forge their own identity. And in many ways the Firm was a unexpectedly original-sounding unit, mixing Page and Rodgers's natural blues tendencies with a spacious eighties new-wave sound and a rubbery rhythm section driven by Franklin's distinctive fretless bass and Slade's sledgehammer groove.

Page even began favoring a new guitar for his band's new sound. He temporarily retired his sunburst Les Paul and took to playing a brown Fender Telecaster modified with a Gene Parsons/Clarence White B-Bender, a device that enabled him to create a pedal-steel effect by mechanically bending his B string up a whole tone. Jimmy had used the guitar sparingly with Led Zeppelin at Knebworth on "Ten Years Gone" and "Hot Dog," and he'd also used it extensively at the ARMS concerts, but his tour with the Firm made it official: Page had a new mistress, and it was serious.

"It took me about a year to come to terms with the B-Bender," Page told writer Steven Rosen in 1986. "I've always liked the idea of how a pedal-steel guitar can change intonations. Then I heard Clarence White use the B-Bender on the Byrds' *Untitled* album, and I thought it was quite amazing. One night I went to see them play, and afterwards I spoke with Gene Parsons, who had co-designed the device with White, and he was kind enough to make a bender for me."

If the Firm and the ARMS Charity Concerts had successfully restarted Page's career, another charity event would kick it back into the stratosphere. Live Aid was a multivenue rock concert held on July 13, 1985. The event, featuring the world's biggest rock stars, was organized by Bob Geldof of the Boomtown Rats and Ultravox's Midge Ure to raise funds for famine relief in Ethiopia. Held simultaneously in Wembley Stadium in London and JFK

Stadium in Philadelphia, the epic concert was televised live and seen by an estimated four hundred million viewers across sixty countries.

Initially, Robert Plant and his new band were approached to play. Plant was touring the U.S. supporting his album *Shaken 'n' Stirred*, and his schedule dovetailed with the Philadelphia show. But after considering the scope and significance of the event, the singer decided to contact Page and bassist John Paul Jones for a one-off Led Zeppelin reunion, albeit without John Bonham. The difficult task of taking his place on drums was assigned to two players: the estimable Tony Thompson, a session drummer best known as a member of the funk band Chic, and superstar Phil Collins of Genesis.

The band took the stage at 8:13 Eastern Standard Time and ripped through "Rock and Roll," "Whole Lotta Love," and "Stairway to Heaven." The players were under-rehearsed and their playing a bit ragged, but—given the near-hysterical response to their appearance and set—no one seemed to notice or care. Fans watching the show on television also reacted in a positive manner by opening their wallets. It was later reported that within an hour of Zeppelin's appearance, pledges on behalf of the beleaguered Ethiopian people had more than doubled.

Inevitably, the exhilaration of Live Aid and the whispers of the Zeppelin reunion it engendered took much of the oomph out of the Firm. Nevertheless, Jimmy and his band returned to the studio and recorded a second album, *Mean Business*, and in March 1986 embarked on a brief tour of the U.S. Soon thereafter, however, the Firm was liquidated.

"The Firm was designed from the beginning to be a two-album project," Page says. "After that, Paul and I agreed it had run its course."

MUSICAL INTERLUDE

———

A CONVERSATION WITH
BAD COMPANY AND
FIRM VOCALIST PAUL RODGERS

I T WAS ALWAYS clear that if Jimmy Page was to collaborate with a vocalist, it would have to be *some* vocalist. Paul Rodgers fit the bill. Contemporaries like Rod Stewart, Pete Townshend, and Freddie Mercury
all sang his praises, and in 2008 Rodgers's raucous, bluesy baritone
earned him a place on *Rolling Stone* magazine's list of the 100 Greatest Singers of All Time.

In 1968, Rodgers rose to fame after forming Free, whose spectacular
1970 hit single "All Right Now" remains a classic-rock staple. During the
mid-seventies, he fronted the FM rock juggernaut Bad Company, who were
signed to Led Zeppelin's Swan Song label. The band racked up one smash
after another, including "Can't Get Enough," "Feel Like Makin' Love,"
"Shooting Star," and, of course, "Bad Company."

Below, Rodgers reflects on the Firm, its music, and his relationship with
Jimmy Page.

What did it mean to Bad Company to be signed by Swan Song?

PAUL RODGERS Led Zeppelin were gods. It was really quite amazing that
they took the time to create Swan Song, which was kind of a talent-scouting

outfit. They gave us a fantastic chance by signing us to their label and allowing Peter Grant to manage us.

**Free and Zeppelin were contemporaries. What was it like to witness their be-
coming, as you say, "gods"?**
We were impressed with their sudden rise. Free had been around for a couple years and Zeppelin just blew right past us. Suddenly they were doing
these big shows and their posters were everywhere.

How did you sign with Swan Song?
After Free broke up, I started putting together Bad Company, with Mick
Ralphs on guitar. I kept thinking we needed a manager. The problem with
Free is we managed ourselves and it only worked up to a point.

Free had a roadie named Clive Coulson who'd left us to work for Peter
Grant and Led Zeppelin. But we were still friendly, and when he came
around he told me about Swan Song and urged me to contact Grant. So I
did, just right out of the blue. We wanted a big manager and a big label, and
Led Zeppelin were the biggest band in the world, so I called Peter to see if
he was interested in working with us. He said, "Well, I'm interested in *you*."
I said, "I come with a band and we're going to be called Bad Company." And
he said, "I don't know about the name, but I'll come down and have a listen."

We arranged for him to attend a rehearsal at the Village Hall where I
used to live, in Surrey. We didn't have a bassist, but we had lots of good
songs, including "Rock Steady" and "Can't Get Enough of Your Love." Anyway, Peter never showed up and we were pretty disappointed. While we were
packing up our gear to leave, suddenly he turned up. He'd been listening
outside of the door because he didn't want to intimidate or affect us in any
way. Fortunately, he liked what he heard.

He said, "You don't know me, and I don't know you, so we won't do any
contracts. For the first three months we'll just work together on a handshake."

Peter was old-school management, and an ex-wrestler, so he knew the
sweaty end of the business, you know? He was very intuitive—a magical

The Firm (from left) Paul Rodgers, Chris Slade, Page, and Tony Franklin, 1985 *(© Neal Preston)*

person, really. I believe he was a Romany Gypsy. He was a huge guy but very gentle as far as we were concerned. We were never afraid of Peter—in fact, we used to insult him constantly. We were terrible! He took it all, and he was really great.

This is an old story, but when we first started our tour in the U.S., the first Bad Company album was ninety-nine on the *Billboard* charts, and when we finished, it was number one. That was pretty good management, I thought.

I definitely had my run-ins with Peter. He didn't want to call the band Bad Company. The record company didn't want it. Nobody wanted it! One of the band members came around to my cottage in order to convince me that it wasn't a good idea to pursue it, but I said, "You know what, it's going to be Bad Company—that's it. That's going to be the name." What's strange is that

I had the same issue with Free. I think that however great an opportunity is, you've still got to stick to your guns.

How did you meet Jimmy?

I met Jimmy at the Swan Song office. I was a little in awe of him because Zeppelin was such a huge monster. What really made Zeppelin impressive was that they were both big commercially and had artistic depth. Despite his reputation for being aloof, I found Jimmy to be very easygoing.

He's still a great friend. He turns up when I do shows in the UK, and when I was recently given the Ivor Novello Award for songwriting he came and sat beside me at my table.

Tell me about the genesis of the Firm.

After several albums and many tours, I left Bad Company. I was so exhausted, I wasn't sure if I ever wanted to tour again. I decided to build a recording studio at home and just make music there. During that period of time, Led Zeppelin lost their dear friend John Bonham, who was the heartbeat of the band. I think Jimmy was at a loss, and he would pop round to my place just to see what I was up to. When he first came around, it had been a while since he played guitar, and all the people around him told me, "Whatever you do when he comes, don't ask him to play guitar."

I thought about that, and as soon as he walked in the studio, I said, "Hey, Jimmy, let's have a jam." It was like, shock . . . horror! But by the end of the evening he was playing and we were jamming. I thought it was crucial. If he was grieving, the thing to do was to keep playing, because that's who he is, he's a musician. So that's how we sort of started doing things.

You were working with Jimmy informally, but that changed.

Jimmy played the ARMS benefit concerts in England and used Steve Winwood as his vocalist. But Steve couldn't do the American tour and they needed somebody to fill that slot. I think it was Eric Clapton's management—who were very involved in organizing those events—who called and said, "We heard you were in the studio with Jimmy. Can you put something together and come out and play?" Jimmy and I told them we were just noodling around and didn't have a band. And they said, "Well, you only have to play half an hour, that's all we need." Basically we ran out of excuses, because we had a half hour of music and they said they could give us a rhythm section. It was such a good cause and Ronnie Lane was such a beautiful guy, we decided to go for it.

I must admit, at that time I was very reluctant to go back out on the road again, but I was excited that we were doing something very worthwhile. At that point there wasn't a plan to form a band, but I think the ARMS concerts inspired Jimmy. He came back from the U.S. very keen to get something together and go out on the road, because he felt it would be good therapy for him. He kept telling me, "We'll put a band together and go out on the road." And I kept answering, "I'm kind of done with the road. I don't really want to go out."

And then we came up with this formula. "We'll do two albums," said Jimmy, "and we'll tour to support those two albums, and that's it." And I said, "Okay." And so that's what we did. We didn't have a contract between us. It was just a handshake agreement.

One of the first songs you guys worked on together was the incredibly ambitious nine-minute "Midnight Moonlight."
Yes, it was. I mean, I like a challenge, and it was the first song that we wrote together. Jimmy brought this huge piece of music, which was originally much longer than nine minutes. I was very sort of loath to say, "Could we make it a bit shorter?" I mean, who am I to tell Jimmy Page that it's got to be a little shorter? But he said, "Yeah, we can cut it back a bit."

How did you go about crafting the melody and writing lyrics for that song? It must have been a tall order.
I think a lot of songwriters would say that you let the music speak to you and see what words come to mind when you hear the music. I just let it speak

to me. I started singing, "The dawning of a new creation." Those chords seemed to be saying that to me and I just identified with those words. I wanted it to be a song of hope for the future.

That is one of Jimmy's lengthiest and most complex compositions, something that is rarely pointed out.
It wasn't very radio-friendly! When you have something that's nine minutes long and is structured like a piece of classical music, you have to really listen and go with it. Perhaps it was too challenging for most people.

The Firm had an incredibly original sound. It was different than Free, Bad Company, and Led Zeppelin. Was it important to you and Jimmy to create something new?
It was a given that we would do something original. Neither of us even considered for a moment doing our old material. It's strange, now that I think about it. I don't know if it was ever discussed; we just didn't go there. It was automatic that if we were going to do something, we'd write together and create something new.

It was rumored that your original choices for the rhythm section were King Crimson's Bill Bruford on drums and studio musician Pino Palladino on fretless bass.
Pino was definitely in the cards. He was always going to join us but was busy with other obligations. We started using Tony Franklin, who also played fretless bass during rehearsals, and soon it became a guessing game of whether Pino was going to join or not. Jimmy just said to me, "Well, Tony has rehearsed with us, he knows all the songs, and he's part of the band now: Shall we go with him?" So we did. I think at the time Pino did not want to go on the road, which I understood.

The fretless bass was something of an unusual choice.
That was Jimmy's idea, and it did bring a different flavor to the band. On something like "Radioactive," we'd start with a simple demo and it would go

to this other place—this Firm sound. Our drummer, Chris Slade, would get this beat going and the bass would do that fretless thing, and it was really great because the rhythm section would elevate the song and really make it take off.

The first Firm album is highly underrated. There are some great songs there, but it is partially undermined by its distinctly eighties production—lots of chorused guitars and overly wet drums.

That was just the sound that was prevalent at that time. You had to have those big wet drums because it was cutting-edge, and that's the way we went. We were just doing what came naturally at the time.

You recorded it at Jimmy's studio, the Sol, right?

Yes, we did it there. It's a lovely studio, really nice. It's on the river and is *very* haunted. It really is. The engineer said, "Oh, you're staying tonight and you're sleeping in *that* room?" And I was like, "Yeah." He goes, "Oh, you'll be all right. See you later." I'll be all right? Why wouldn't I be all right?

And so I'm sleeping when, all of a sudden, in the middle of the night, this great bird flew across my face and went straight into the cupboard. I went, "Holy shit!" And I put on the light and went to the cupboard; of course, there's nothing there at all. Then I thought the window must be open, so I go over to the window and it's triple glazed—nothing at all could have possibly flown in. Some other things happened. It was all very strange. Let's put it this way: I didn't stay there anymore.

You play a lot of guitar on the album.

Not a lot of people know this, but I actually played the solo on "Radioactive." It was a finger exercise that British blues pioneer Alexis Korner taught me years and years earlier, and it was so robotic and weird that I wanted it to be part of the song. When I think back, it was actually kind of ballsy of me to say to Jimmy Page, "I'll be doing the solo on this one, Jim." But he was fine with it, and he contributed some great chords.

You've worked with many great guitar players, including Paul Kossoff in Free, Mick Ralphs in Bad Company, and Brian May in Queen. What is it about Jimmy that makes him special?

He has an incredible, almost mathematical mind. He can come up with the most amazing chords, and he inserts them in places you wouldn't expect—go back and listen to the chord voicings in "All the King's Men." And he could actually lift the entire band with just a guitar solo. His sound was a sound you could almost taste. I used to just stand on the stage and go, "Wow!" You could almost feel it. I think, technically, he's probably one of the greatest guitar players in the world—him and Jeff Beck.

Jimmy was also very open. At one point he said to me, "Let's do a cover song. What song would you like to do? Any song at all." And I said, "I've always wanted to do the Righteous Brothers' 'You've Lost That Loving Feeling.' " We did our own version of it, and I thought it was incredibly generous of Jimmy to do something completely off the wall like that.

Overall, I think his approach to the Firm was very different, wasn't it? It was very atmospheric. The Firm didn't have the heaviness of Zeppelin or Bad Company.

How did working with him compare with your experience with other musicians? Did you find his methods to be unusual?

He's a bit of a production genius. I pretty much left the studio stuff to him, to be honest with you, because he was just great at it. I don't want to say that he's just technical, because he's got a great feel, too, but he has more studio knowledge than most musicians—certainly more than me. I had to struggle a little bit to keep up. The way I do music is to just feel it.

Did the songs and music come together relatively quickly? I know Jimmy likes to be efficient in the studio.

Oh, so do I. I hate to waste studio time. I've always had that approach, and Jimmy concurred: You do all the legwork at rehearsal and you don't waste studio time, because it's usually expensive. I hate that kind of waste. Also, I like to do the hard work in the rehearsal room so that when you come into the studio you know exactly what you're going to be doing and you can focus on the performance. That makes it much easier to get the kind of quality you want.

I've heard that when you sing you like to get the vocal down in one or two takes.
Well, yeah. If you do a hundred takes, you'll lose something. You might get the vocal perfectly right, but the energy and joy goes out of it. You're always trying to find something inspired in a performance, and it's a one-shot thing.

It's almost like acting—you have to be in the moment for it to ring true.
That's exactly right. I think you have to put yourself into the song. I learned that from listening to people like Otis Redding and Wilson Pickett, and blues guys like John Lee Hooker, B. B. King, and Albert King. They sang those songs like they meant it, and that's the only way to do it. And that's what I mean. You want to master the song in rehearsal so you can forget about the technicalities when it comes time to really deliver the song in the studio. Just do it like you mean it. I think that's the way it should be for all musicians.

What was the best thing about the Firm?
For me, it was that it got Jimmy up running, rock and rolling, and back into music. That was my whole purpose in the Firm. And I was so happy to see him back in form. On a musical level, I enjoyed "Satisfaction Guaranteed," "Radioactive," "Midnight Moonlight," "Lady"—I enjoyed some of the things we did creatively. And we did some *great* shows.

Fans and music critics alike find that there is a mystique to Jimmy that transcends even his musicianship. Was he just a regular guy to you?
Jimmy was a very regular guy in some ways, but he always had this extra dimension to him. Almost a wizardry. He's an alchemist in many respects. He puts a drop of magick in whatever he does.

When the Firm came to an end after two records, did you feel like you had done what you'd set out to do?

Well, I had accomplished what I wanted to do, and that was to see Jimmy happy. He was in great, great shape when we ended.

And why was that important for you?

It was very important for me because I had lost Free guitarist Paul Kossoff to drugs and depression. I've always regretted not being able to do anything for him. And I was concerned that we might lose Jimmy.

10

PAGE RECORDS HIS ONE AND ONLY SOLO
ALBUM, REVISITS LED ZEPPELIN, AND RECORDS
A CONTROVERSIAL ALBUM WITH HAIR-METAL
PINUP DAVID COVERDALE.

David Coverdale and Page, 1993 (© Ross Halfin)

"I STILL HAD A LOT TO OFFER
AND SAY MUSICALLY . . ."

———————

IN DECEMBER 1986, Page married twenty-four-year-old Louisiana native Patricia Ecker. They met in New Orleans, where she was working as a waitress in the French Quarter. Jimmy would later say the connection was immediate, and they would stay together until 1995, having a son, James Patrick Page III, born on April 26, 1988. The couple moved to Jimmy's house in Windsor, where he began to formulate *Outrider*, his first and only solo album. In the album's initial stages, Page considered making it a double-album set, with each of the four sides focusing on a different aspect of his playing.

Unfortunately, early in the composing process, several of the demos Page had selected to work from were stolen from his house, along with demo tapes of Led Zeppelin and other personal belongings. The loss of the demos caused the guitarist to rethink his original idea. But there were other reasons as well.

As Page told journalist Bud Scoppa in 1988, "Because I was shaping *Outrider* as I went along, I put more work into it than any other album I've ever worked on. Consequently, I didn't fancy doing a double—it would've been a masochistic task."

That Page put an enormous amount of thought into this rich and varied work is evident. He stated that for every solo and overdub on the album, he often tried as many as four completely different approaches before deciding which to use. In many ways, Jimmy's playing on *Outrider* felt like the ultimate expression of a new and more personal approach that he began exploring as far back as *Physical Graffiti*. Looser yet rhythmically more angular and complex than his early work with Zeppelin, Page's "guitar voice" was distinctly different than the one heard on *Led Zeppelin IV* or *II*.

Released on June 19, 1988, *Outrider* is a fairly comprehensive survey of Page's late-eighties guitar style. Many of the tracks are doubled with additional layers of overdubs and harmony guitars that beckon the listener to go deeper into the mix. Even the more straightforward tracks, like the big and beefy "Wasting My Time" and "Wanna Make Love," with blues belter John Miles on vocals, are spiked with tricky, off-kilter rhythms and ricocheting fills, intriguing textures, and sharp, explosive solos.

The album was rounded out by three wonderful instrumentals, a couple of intense minor blues sung by Chris Farlowe, and an additional rocker featuring Robert Plant. Perhaps the presence of three different vocalists made it difficult for critics to embrace *Outrider* as a coherent masterwork, but the subsequent tour was met with almost universal acclaim. Unlike his concerts with the Firm, Page's 1988 *Outrider* shows—featuring Miles on vocals, John Bonham's son, Jason, on drums, and Durban Laverde on bass—were overviews of his entire career and included early efforts like the Yardbirds' arrangement of "Train Kept A-Rollin'," Zeppelin-era gems such as "Over the Hills and Far Away" and "Custard Pie," and more recent selections from *Death Wish II*, *Outrider*, and his albums with the Firm.

The tour was a precursor to Page's next huge undertaking: a long-overdue reassessment of the Zeppelin catalog. In the decade since the band disbanded, vinyl had all but disappeared, and the entire Led Zeppelin oeuvre had been hastily transferred to compact disc. "Atlantic put out the catalog in CD form, and I wasn't involved," Page grouses. "I heard horrific stories of how they were 'mastered,' if you can call it that. I don't think they came up to scratch. Like, they cut off a cough at the end of 'In My Time of Dying,' which would irritate any Zeppelin fan."

Page and recording engineer George Marino spent long days and nights at New York's Sterling Sound Studios, sprucing up Zep's best. The result was 1990's *Led Zeppelin,* an elaborately packaged four-disc set on Atlantic Records. But in digging up long-vaulted master tapes and running them through modern EQ systems, not to mention personally supervising song selection and sequencing, Page was after more than a profitable stroll down memory lane: "I absolutely wanted a new look at the band and its music."

The end result was a vast improvement in overall sound quality that compared favorably with the standards of the time. Marino said the entire process wasn't easy and demanded a number of difficult aesthetic and technical decisions.

"Jimmy spent months searching for the original studio tapes," the engineer told journalist Joe Bosso in 1991. "As much as possible, he tried to make sure that our source tape was the original master tape. You can't imagine how much detective work was required. There were tapes at Atlantic, tapes in Jimmy's home, tapes in different studios around the world. Even with an important artist like Jimmy Page, there was a good chance that tapes from 1969 were vaulted, with no one knowing where they were. Jimmy did all of that, searching high and low.

"When we finally assembled the masters, we did have to correct some sound loss created by the passage of time. But generally, what we decided to do was approach each song on a music and energy level—if we came up with an equalization that made the tune really cook but caused some ill effects— a little noise or distortion—we ignored it. The fact that it was coming out on CD didn't mean that everything had to be perfect and pristine.

"I thought the project was going to be a nightmare, but Jimmy really made the whole thing quite fun. He knew there were so many fans out there that would go crazy for this record, and he really wanted to put it out with a lot of integrity. And it was great to see him get into the music."

The large and lushly packaged boxed set could easily have served as Page's grand finale—a monument to a momentous career. Instead, the project reenergized him and, at the suggestion of A&R executive John Kalodner, Jimmy teamed up with Whitesnake singer and Geffen Records labelmate David Coverdale in 1991 to start working on an album.

"The success of the Led Zeppelin box set could've been intimidating, but it reminded me that I still had a lot to offer and say musically," Page says. "David was the key element. I hadn't had a writing partnership like that since the early days of Zeppelin, and it was the most focused I had been in a decade."

While Jimmy was enthusiastic, the media was quick to pounce on the project as being beneath the guitarist, dismissing the flamboyant Coverdale as nothing more than a Robert Plant wannabe. Page flicked away the criticism like so many Lucky Strike stubs: "We have the same foot size and we smoke the same cigarettes. That's what I call a partnership!"

Coverdale was less flippant. The extroverted vocalist readily admitted that he had some image problems, but he saw his collaboration with Page as a way to distance himself from his recent past that included starring in over-the-top, bump-and-grind MTV videos like "Here I Go Again." "I was incredibly disenchanted with being in Whitesnake because of the peripheral aspects of being a musician—the videos, the press, the makeup—were suddenly becoming more important than writing and singing," Coverdale says. "It had degenerated into, you know, 'Excuse me while I slip into something more uncomfortable,'" referring to the tight spandex clothes that became the costume de rigueur for metal bands in the eighties.

Despite the skepticism, the *Coverdale and Page* album released in March of 1993 was a hit both in the U.S. and the UK, achieving platinum status. It was a short-lived collaboration, but it served to inspire Page and Plant to hook up one short year later.

CONVERSATION

Q:

HOW DID THE *COVERDALE AND PAGE* PROJECT COME TOGETHER?

After the release of the Led Zeppelin box, there was a good nine or ten months of inactivity, although I was champing at the bit to play again. During that time, there was a feeling among my management and those of John Paul Jones and Robert Plant that we were going to reunite in some shape or form. We weren't really sure whether we were going to plan the "legendary tour that never happened" or record a new album, or do both.

Unfortunately, it wasn't to be. So after that frustrating interlude, my next option was to consider recording another solo album. I began to search for a singer. I waded through scores of cassettes, but none of them even gave me the inspiration to get on the phone and call a manager.

Then my own manager at the time called and asked whether I had ever considered working with David. I said, "That's interesting. He's a damn fine singer. Let's see how we get on socially." I figured that if we couldn't carry on a conversation, there would be no way we'd be able to write together. As it turned out, we got along famously.

The next step was to see how well we could write together. Just because we were two "heavyweight figures" didn't necessarily mean that we'd be able to create the sort of magic needed for us to join our two names. We decided to give it two weeks. I figured that if we were lucky, we would be able to put together about four songs. As it turned out, we clicked right away and music came pouring out.

Were all the ideas new?
No—I did use one riff that I had in the back of my mind for quite some time. The acoustic lick that opens "Shake My Tree" was something I'd originally presented to Led Zeppelin during the *In Through the Out Door* sessions. At the time there wasn't any sort of mass enthusiasm for it. No one except for Bonzo really seemed to understand what to do with it, so I filed it away.

I decided to pull it back out, and David grasped it instantly. He came out with the lyric on the spot and we started sparking off each other until we came up with the total construction.

Robert went public with his disapproval of the collaboration.

He took a lot of snipes at me. Even when I went to the U.S. to do publicity for the *Outrider* album, all I heard was, "Robert said this" and "Robert said that." It was really bothersome. I continually had to say, "Aren't we supposed to be talking about *Outrider?*"

After you and David developed some sense that you could work together, what happened next?

Our ideas were coming so fast that we just needed something to capture them quickly. We started using this rather primitive fifty-dollar Radio Shack cassette recorder that sort of became our lucky charm. All we had was a little Vox amp, an acoustic guitar, a tape that had some drum rhythms on it, and the recorder.

After we had written about five or six songs, we brought in bassist Ricky Phillips and Heart drummer Denny Carmassi and recorded demos on an Akai twelve-track machine. When we completed the demos, David and I went back to the Radio Shack recorder, wrote some more, and repeated the process with the Akai. The album's rhythm tracks were recorded at Vancouver's Little Mountain Studio, which had a great reputation for capturing great acoustic drum sounds, and we finished the vocals and overdubs at Criteria Studios in Miami.

In terms of overdubs and texture, the album represents the greatest realization of your "guitar army" concept.

The general signal from the guitar was usually split between two or three completely different amp setups. The setups were then mixed together to make up the sound. I always wanted to take this approach to recording but never had enough tracks. This time I had seventy-two tracks to play with, so I took advantage of it.

The motivation behind the project was to take our time and maintain the quality. I wanted to present the best I could get out of myself. It was the best I'd played since the days of Led Zeppelin.

How were the rhythm tracks recorded?

Like the lead and overdub tracks, they were recorded using composite systems, but there were some important differences. First, all the rhythm tracks were played live with the rest of the musicians. But also, in addition to the various changes in guitars, amps, and effects, we also always recorded a direct feed—a dry signal from the guitar that went directly into the board.

When you play live, certain push beats are crucial to maintaining the natural feeling of the song. By the time you start layering and adding overdubs, you may find that the original rhythm guitar starts getting weak and needs beefing up. When that became the case on the album, I took the direct input signal, ran it through an amplifier, and re-recorded it to reinforce the original rhythm part. I didn't have to do that all the time, only sometimes.

Did you record in the booth or in the room with the amp?

Definitely in the room with the amp and the speakers. The interplay between the guitar's pickups and the amplification is an important component of my sound. That would be lost if I stood in the control booth. I like to get my sound so that it borders just on the edge of feedback. The only drawback to staying in the recording space is that you have to wear headphones. I don't particularly enjoy them, but they are a necessary evil.

Did you ever double your parts manually? That is, did you ever reinforce your parts by playing them more than once with different amps and instruments?

Yes. "Over Now" is a good example of that. But there's an interesting story attached to that song. The band as a whole did a couple of takes, and after we sat down and decided which performance felt the best, I was going to use a different setup to double the whole rhythm track. But first, just to see exactly how on the ball my playing was, I decided to compare the performance of one of the rejected takes to my performance that we decided to keep. It turned out that they were almost identical—it was uncanny. I was right on the mark both times. That was very encouraging to me!

While the album was done on very state-of-the-art equipment, you stayed away from digital rack effects.

That's true. I did use a DigiTech Legend 21 rack unit, but generally I stayed with floor pedals. I pioneered the use of pedals, so why not? Basically it was just easy for me to sit back and think, I'll put this, this, and this together and go for it. Probably my favorite effect was the one used on "Over Now." After David sings, "I release the dogs of war," you hear this growl. I produced that by running my purple B-Bender Les Paul through an early-sixties Vox wah, a DigiTech Whammy Pedal set "deep," an old Octavia, and one of my old one-hundred-watt Marshall Super Leads, which I used with Zeppelin.

You played harmonica really well on "Pride and Joy."

I played harmonica years back when I was doing session work. I hadn't really played in about twenty years, so it was fun. But it took me about two hours to recover. I blew so hard, I saw stars!

Were you in any way driven by a feeling that you had to live up to your legacy?

That's never been a motivating force. Naturally, I have incredibly high standards for myself and I'm my own worst critic. I know when I'm not playing well.

Looking back today, what's your assessment of the *Coverdale and Page* album?

There was no BS in any respect or in how we executed. I wanted to show that I was still alive and kicking, and in that regard it was a total success.

MUSICAL INTERLUDE

AN INVENTORY OF JIMMY PAGE'S
PRIMARY GUITARS,
AMPS, AND EFFECTS

WHILE MOST GUITARISTS TEND TO PLAY THE FIELD WHEN IT
COMES TO THEIR INSTRUMENTS, PAGE CULTIVATED A RATHER
SMALL AND POWERFUL SIX-STRING HAREM. HE HAS CERTAINLY
HAD A NUMBER OF MINOR FLIRTATIONS OVER THE YEARS,
BUT THE FOLLOWING GUITARS, AMPS, AND EFFECTS ARE HIS
MOST CONSISTENT AND PHOTOGRAPHED COMPANIONS.

1960 GIBSON LES PAUL CUSTOM

This "Black Beauty" was purchased new in 1962 for £185 and was used on much of Page's session work from 1963 to 1966. Perhaps the most distinctive features of the Custom were its huge Bigsby vibrato tailpiece and three pickups, versus the two found on other Les Paul models. In simplest terms, the extra pickup offered Page more colors and tones, from bright, shiny trebles to silky-smooth lows and mid-range tones. During his session years, he never knew from day to day what type of music he would be required to play, so having many sound options at his fingertips was crucial to his success and livelihood. In this way, the tuxedo-black instrument was the ideal instrument.

Always interested in pushing the sonic envelope, Page made additional modifications to the guitar. While in Led Zeppelin, Page and American guitarist/technician Joe Jammer installed a unique system of three toggle switches that had other players doing double takes.

"It originally had a single toggle switch that only allowed Jimmy to choose between each pickup," Jammer said. "I took out the single switch and put in three on/off switches. The system allowed him to use any combination of pickups on or off."

Page loved the guitar so much that he rarely took it on the road for fear of doing it harm. But things were going so well with Zeppelin that he had a change of heart and toured with it from January through April, 1970. Unfortunately, disaster struck. "It was stolen off the truck at the airport while we were heading to Canada, and it never arrived at the other end," Page justifiably moans. He famously placed an ad in *Rolling Stone* magazine offering a reward for its return, but to no avail. He has always referred to the instrument as "the one that got away."

1959 TELECASTER ("THE DRAGON TELE")

Given to Page by Jeff Beck in 1966, it was Jimmy's main ax while he was in the Yardbirds and was used live and in the studio with Led Zeppelin from 1968 to May 1969. It was originally painted white, but Page added eight reflective circles in spring 1967, and later that year he stripped it down and personally hand-painted it with a dragon theme. It is the primary guitar heard on *Led Zeppelin I*. He later used the instrument to record the solo to "Stairway to Heaven."

1959 GIBSON LES PAUL STANDARD ("NUMBER ONE")

"It's my mistress, it's my wife," Page says of his main Les Paul. "It's absolutely irreplaceable."

It's hard to say what makes love at first sight. While hardly a "looker" by some flame-top standards, Page's Number One makes up for its lack of visual punch with its outstanding sound. And something about it must have just felt right.

The honey sunburst 1959 guitar—acquired from then–James Gang guitarist and future Eagle Joe Walsh in 1969 for $500 and heard on the bulk of Zeppelin's output from then on—is distinctive in several ways. The slim, elliptical neck profile (and lack of serial number on the headstock) is due to the fact that Walsh had it sanded down before selling to Jimmy. In addition, Page had the guitar fitted with Grover tuners, which he became familiar with through his 1960 Les Paul Custom.

"The Grovers are more sensitive," Page explains, "and, boy, they've held up from all those days—that says it all. In a three-piece like Zeppelin, you couldn't have slipping machine heads. It's one of those guitars that was meant to come my way. Joe Walsh absolutely insisted I buy it. And he was right."

Page loved his guitar, but he had no problem sending it in for a nip and tuck. In the eighties, he had Steve Hoyland, his studio maintenance engineer, enhance the guitar by adding push/pull tone pots that allowed for the sort of reverse-phase sound Page associated with "a close approximation of the sound Peter Green would get, and certainly B. B. King."

1959 GIBSON LES PAUL STANDARD ("NUMBER TWO")

Every guitarist needs a good backup instrument in case of technical emergencies or tuning problems onstage or in the studio. After a lengthy search for a guitar worthy enough to play second fiddle to his Number One, he finally purchased this dusky amber-brown Les Paul in 1973.

Number Two was essentially original when he acquired it, but as with his main instrument, it was subjected to several key modifications. First, he had the neck sanded down to generally match the feel of his Number One. In the early eighties he decided to customize it so he could "explore the full range of what the two humbucking pickups had to offer." Page designed a sophisticated switching system for coil-splitting, series/parallel, and phase-reverse options for both pickups. The result comprised a push/pull pot on each of the guitar's four standard controls, plus two push-button switches beneath the pickguard, allowing him to reproduce an incredibly wide range of tones, from the thinnest highs to the fattest lows.

HARMONY SOVEREIGN H1260 FLATTOP

Page can't remember exactly where or when he bought his Harmony acoustic, but he ventures that it might have been at a local music store around the time of the Yardbirds. More upscale acoustics, like Martin and Gibson, were not readily available in England at that time, so he grabbed this perfectly respectable jumbo flattop. With its huge bass response and clear, sharp high end, the Sovereign served Page well, appearing on such studio recordings

Robert Plant with Page and his EDS-1275 6/12 double neck (© *Jim Marshall Photography LLC*)

as "Babe I'm Gonna Leave You," "Ramble On," "Friends," and, of course, "Stairway to Heaven."

1971 GIBSON EDS-1275 6/12 DOUBLE-NECK

While it was rarely used in the studio, the EDS double-neck guitar is certainly one of Page's most iconic instruments. The object of countless photographs, the unusual two-guitars-in-one construction was both visual and eminently practical. In many of Page's greatest compositions, he often switches between six- and twelve-string guitars midsong. While this is easily accomplished in the studio, such lightning-fast changes are typically impossible in live performance. Using the Gibson with two separate necks—one strung with twelve strings and the other with six—allowed him to make these transitions in concert with style and ease. Though he originally procured the guitar for Zep's live version of "Stairway to Heaven," Page soon began using it onstage for "The Song Remains the Same" and others.

1965 FENDER ELECTRIC XII

So if Jimmy didn't use the Gibson double-neck in the studio, what twelve-string electric did he use for songs like "Thank You," "The Song Remains the Same," "When the Levee Breaks," and "Stairway to Heaven"? According to Page, his guitar of choice was a sunburst Fender Electric XII, which he acquired while he was in the Yardbirds.

The XII has rarely been photographed or seen, but it made a rather dramatic public appearance in the spring of 2009 when Jimmy pulled it out of mothballs to perform "Beck's Bolero" and "Immigrant Song" at the Rock and Roll Hall of Fame dinner with inductee Jeff Beck.

1961 DANELECTRO 3021

Constructed of Masonite and poplar, the Danelectro was often dismissed as a cheap economy guitar, but that didn't stop Page from seeing its potential. Drawn to the instrument's rather appealing hollow twang, which sounded somewhat like an amplified acoustic guitar, he used it live for songs in sitar-like D A D G A D tuning, including the "White Summer/Black Mountain Side" medley and "Kashmir." He also used the instrument for slide songs such as "In My Time of Dying."

In 1982, the original Danelectro stainless-steel bridge was replaced with a Leo Quan Badass bridge with individual adjustable saddles for more precise tuning.

1953 FENDER TELECASTER

The only guitar to ever threaten the primacy of Page's Number One Les Paul. Bought in November 1975, the brown Tele was used for "Hot Dog" and "Ten Years Gone" on the 1977 U.S. Zeppelin tour. Two years later, for the 1979 Knebworth shows, Page switched the maple neck for the rosewood neck from the 1959 "Dragon" Tele.

This Tele is perhaps most famous for being retrofitted with the Gene Parsons/Clarence White B-Bender, allowing Page to bend the guitar's B-string up a whole tone (two frets) to C-sharp to approximate the sound of a pedal-steel guitar. Page, however, used the device to create his own quirky textures, a nice example being the languorous, sliding solo on Robert Plant's solo version of "Sea of Love."

In terms of tone, the biting brown Telecaster couldn't have been more different from Page's lush-sounding Number One Les Paul. But perhaps that was the point. After Led Zeppelin folded, Page was interested in carving a new path and used the Tele as his main guitar in the 1983 ARMS tour and continued to focus on it throughout much of the eighties in the Firm and on the *Outrider* tour.

AMPS & EFFECTS

It's a bit of a fool's game to speculate which amps were used on what songs in the Zeppelin catalog. Whether it's because he can't remember or prefers to keep people guessing, Jimmy rarely talks about his studio amplification. When pressed on the subject, he demurs, saying that whenever he's mentioned using a particular brand, they all get bought, making it difficult for him to find suitable replacements.

That said, we do know a few things for certain. In the sixties, a Vox AC30 could be found in most British studios and stages. In his Yardbird days, Jimmy, like many of his peers, used a version of this amp featuring a "Top Boost" (or "Brilliance") feature that added some extra snap, crackle, and pop to the sound. This thirty-watt combo would suffice early on, but larger venues prompted a need for more volume, and Page started using Arbiter Power One Hundred heads and four-by-twelve speaker cabinets on his early tours with Zeppelin. This early period also saw the use of a Vox UL-4120 120-watt amplifier.

Touring the United States in the sixties had its challenges, and the expense of shipping equipment overseas was one of them. Amplifiers made in the United States were used on these early tours, and Page could be seen onstage with a hodgepodge of Rickenbacker, Univox, and Fender amps and cabinets.

Sometimes off-the-shelf equipment could not provide the performance Jimmy needed, so modifications and/or custom-made units were necessary. The first was the Hiwatt Custom 100 "Jimmy Page" amplifier. This unit incorporated a boost foot switch that eliminated the need for a fuzz tone and was used on tours from 1969 to 1971, until his final change to a 1959 Marshall Super Lead, the amplifier he is most closely associated with.

Effect pedals provided new sounds and textures, and Jimmy employed a few specific units throughout the years. Fuzz tones and wah pedals were the first types of guitar effects on the market, and Jimmy favored the Sola Sound Tone Bender and the Vox grey wah pedal. Tape-delay units were another new invention of the time, and he experimented with several different models until he settled on a Maestro Echoplex EP-3.

Creative combinations of these effects could result in never-before-heard sounds, the most exotic being the union of the Echoplex and his Sonic Wave theremin, an electronic musical instrument that is controlled by the proximity of the player's hands to a pair of antennae, resulting in high-pitched, banshee-like sounds. The classic example of this eerie one-two punch can be heard in the middle section of both the studio and live versions of "Whole Lotta Love."

[CHAPTER]

11

PAGE REUNITES WITH PLANT, TOURS WITH
THE BLACK CROWES, PERFORMS ONE MAGICAL
EVENING AS LED ZEPPELIN, BECOMES A HERO
IN CHINA, AND MAPS A NEW FUTURE FOR
HIMSELF VIA THE INTERNET.

The Led Zeppelin reunion, two minutes before the show, 2007 (© Ross Halfin)

"WE'RE OLDER AND WISER . . ."

R ECORD EXECUTIVES, MANAGERS, journalists, and fans continued to pester Jimmy Page and Robert Plant to reunite Led Zeppelin, but by the early nineties most assumed that the duo's dancing days were over. However, in 1993 Plant was offered an invitation to perform on the hugely popular *MTV Unplugged* series, a program that presented superstars including Eric Clapton, Paul McCartney, Bob Dylan, and Bruce Springsteen playing their greatest hits in an acoustic setting.

MTV wanted Plant to dip into his Zeppelin past, but the singer couldn't imagine doing it without Jimmy. The timing seemed right. Page and Plant had performed with each other for pleasure over the previous several years, and whatever tensions or differences that had existed during the last days of Zeppelin had drifted away like rings of smoke through the trees.

As Page told writer Charles Shaar Murray in 2004, "I was on my way to Los Angeles to rehearse with David Coverdale for a tour of Japan, when Robert's management asked me to pop in and see him in Boston. Robert said, 'I've been asked by MTV to do *Unplugged*, and I'd really like to do it with you.' I agreed. It was a great experience, because it gave us a chance to revisit some numbers and use that same picture with a very, very different frame."

The chemistry between the two musicians was fascinating and undeniable. During the eighties, both had made interesting music as solo artists—interesting, but far from approaching the grandeur of their collaborations in Led Zeppelin. When they put their heads together, they were able to realize a combined vision that was greater than the considerable sum of its parts. Where most rockers who appeared on *Unplugged* saw the show as an opportunity to present their greatest work in an otherwise unrealizable intimate setting, Plant and Page, characteristically, took a far bolder and audacious approach. Their reunion was not going to be a quiet evening of exquisitely crafted nostalgia; it would be a spectacle.

Throughout the first half of 1994 they pulled together a virtual battalion of musicians and arrangers from all over the world to rethink, reinvent, and revisit the music of Led Zeppelin. The rough idea was to highlight and expand upon the Celtic, Middle Eastern, North African, and African American elements that informed the band's back catalog. Ultimately, they enlisted a seven-piece rock-folk band featuring hurdy-gurdy, mandolin, bodhran, and banjo; the eleven-piece Egyptian Ensemble; four musicians from Morocco; and twenty-eight string players from the London Metropolitan Orchestra.

To put an exclamation point on their multicultural *Unplugged* extravaganza, Page and Plant filmed special performance segments in Marrakech, Morocco, and Snowdonia, North Wales, where they had written some of the acoustic songs that appear on *Led Zeppelin III*. For Page, the Moroccan experience, filmed in August 1994, proved to be particularly inspiring. He and Plant, accompanied by Gnawa musicians from the region, performed several new compositions in the streets of Marrakech. The locals watched with curiosity and delight as the British duo performed North African–spiced originals like "City Don't Cry," "Wha Wha," and "Yallah."

"When Robert and I worked with the local Gnawa trance musicians in '94, it was just as stimulating as working with Bombay musicians in 1972," Page says. "The Gnawa musicians play at celebrations and weddings, and then they'll go in a house and exorcise it of demons. They play a very spiritual role in a very different cultural environment. They really didn't know our music at all—and didn't really care about it. But our desire was to simply make a connection so that everyone would go away from the encounter and say, 'Yeah, it was cool. I remember playing with those guys from England and that was really interesting.' "

A short couple of weeks later, Page and Plant reconvened at London TV Studios with more than forty musicians to film the bulk of what would become *No Quarter: Jimmy Page and Robert Plant Unledded.* Months before the taping, keyboardist Ed Shearmur and Egyptian percussionist Hossam Ramzy were given the enormous task of crafting the ambitious orchestral and ensemble arrangements that would fuse the diverse musical styles into a coherent and powerful whole.

"Ed Shearmur and I spent a lot of time together trying various ideas and sounds, rhythms and grooves, and then we would test them at his home studio," Ramzy told journalist Neil Davis. "Ed would present these ideas to Robert and Jimmy. If they liked something, we'd bring the Egyptian musicians to a rehearsal with the core rock band and see if they gelled."

During this process, the arrangers decided the best way to meld what appeared to be impossibly diverse styles together was to present them side by side instead of superimposing them. From that point on, they searched for clean pivot points between the Egyptian rhythms and the rhythms of Led Zeppelin, and used them with great effect on the arrangements of "Kashmir," "Friends," and "Four Sticks." The dynamic renditions showcased the flexibility of the Zeppelin source material without diminishing the power of either interpretation.

The ninety-minute *No Quarter: Jimmy Page and Robert Plant Unledded* aired on MTV in the United States on October 12, 1994, and was a critical and ratings sensation, eclipsing even Eric Clapton's much-ballyhooed comeback performance of 1992. The triumph of *Unledded* encouraged Jimmy and Robert to continue their collaboration, which they did with *Walking into Clarksdale.* Released in 1998, it was the duo's first full album of all-new songs since Led Zeppelin's 1979 finale, *In Through the Out Door.* Featuring the *No Quarter* rhythm section of Michael Lee on drums and Charlie Jones on bass, the album was a surprisingly stripped-down affair, as spare and understated as *Unledded* was extravagantly baroque. Led Zeppelin always took pride in doing the unexpected, and in that tradition Page and Plant surprised the rock world by hiring iconoclastic punk-rock producer/engineer Steve Albini to record what would become *Walking to Clarksdale*.

Albini built his reputation recording no-frills albums with cutting-edge bands like the Pixies, Jesus Lizard, and Nirvana, and while he appeared to be an unusual choice for classic rockers like Page and Plant, the trio's philosophies were remarkably in sync. Like Page in his Zeppelin days, Albini recorded quickly and inexpensively, with an eye toward capturing great performances. He believed all you needed to make a great record was a band that could play the material and an engineer with the ability to select and place microphones.

"I pretty much do things the way Jimmy and Robert have always done them," Albini told *Guitar World* magazine. "In the days before there was so much dependence on technology in making records, there was a very organic performance aspect to everything that was recorded. I have to capture that kind of thing, as a matter of course, for underground rock bands that don't have much money to spend on studio time. And that was actually something that Jimmy and Robert were looking for."

Page explains further: "I always saw *Walking into Clarksdale* as a collection of songs and moods that, hopefully, presented a musical landscape. 'Shining in the Light' is kind of the access point to this whole landscape, with high peaks and mountains and smoky valleys. It's an atmospheric record, really, and something with a lot of information, even though we knew it was going to be very minimalistic. That was the direction we wanted to take after the *No Quarter* project. But because that had so many musicians on it, it was hard to hear the subtleties of what I was doing. So it was back to what we knew best, which is writing songs for bass, drums, guitar, and voice."

One could see how a more direct approach would appeal to the guitarist. On *Outrider* and his collaboration with David Coverdale, he explored the outer limits of multitracking and modern digital recording. On *Clarksdale*, Page was interested in returning to the source of his great powers. Even the album title had powerfully rootsy connotations. Clarksdale, Mississippi, where artists such as Robert Johnson, Son House, and Muddy Waters worked and played, is regarded by many to have been the epicenter of Delta blues—the music that inspired a generation of British rockers. Conceptually, it served as a powerful symbol of Page and Plant's desire to get back to where they once belonged.

Much like Led Zeppelin's *Presence*, Page wanted *Clarksdale* to be "a performance album," where every note played was in its place "to mean something." No embellishments were needed.

As to his relationship with Plant while they worked on the project, Page unabashedly expressed his fondness for his old partner. "We spark off each other in such a brilliant way," he says. "I'd missed Robert's voice and the working relationship. He'd certainly missed my guitar and that very aspect of inspiring each other. It was really fortunate that we still had that ability to conjure up that spirit after fourteen years. The chemistry was immediate."

Immediate it was, but not destined to last for long. Much like the Firm, the Page and Plant experiment produced two very interesting albums and a couple of tours, and then dissolved into the ether as quickly as it materialized.

For the next decade, Page and Plant went their separate ways, drifting from one interesting diversion to another like two feathers in the wind. As the singer of his own songs, Plant had the distinct advantage when it came to putting together bands and various recording projects. In 2007, he hit commercial pay dirt with the Grammy-winning *Raising Sand*, an Americana-inspired album that featured the contemporary bluegrass star Alison Krauss.

Page, for his part, struggled to find musical partners worthy of his attention and focused instead on carefully selected short-term projects. He occasionally returned to his beloved Led Zeppelin, curating and producing the *Led Zeppelin* DVD and the live CD *How the West Was Won*, both released in 2003. He also oversaw the meticulous resuscitation of Zeppelin's classic 1976 concert film, *The Song Remains the Same*, re-released in 2007.

Among the more intriguing one-off projects Page took on was his 1998 collaboration with hip-hop impresario Sean Combs, aka Puff Daddy and P. Diddy. Produced by Rage Against the Machine guitarist Tom Morello for the blockbuster *Godzilla* movie soundtrack, "Come with Me" featured Diddy rapping over Page's guitar on a heavily orchestrated arrangement of Zeppelin's "Kashmir."

"It was a lot of fun," Page remembers. "I got a call asking me if I would be interested in collaborating with Diddy on a remake of 'Kashmir.' He didn't want to sample it and wanted to know if I would play my guitar parts with a live band. All this sounded good.

"But it became even more interesting to me when it was explained that he wanted to record his parts in Los Angeles and link to me in London via satellite. He called me up and said that at some point in the song he wanted to add a modulation. I explained to him that because the guitar is in an open tuning, I prefer to play it in D. And then I suggested that we modulate to E. He paused for a moment and said, 'I don't know nothing about no D's and E's,' which I thought was a great answer!

"After we recorded the track, he told me he wanted to put an orchestra on it, and I said, 'Great, good luck.' It turns out he overdubbed two orchestras on it to create this massive stereo effect. I mean, the guy may not know anything about D's and E's, but he's got a fantastic imagination.

"The final product was epic, and I really enjoyed doing it. When we performed 'Come with Me' live on *Saturday Night Live*, he impressed me again. He kept changing the arrangement all through sound check and dress rehearsals. I had to really concentrate and keep counting, to come in at the right places. I thought, 'He's never going to remember all these changes. He'll never get this right.' But he was right on the nail every time. So you've got to give him his due for that.

"What's interesting is people didn't understand quite why I worked with Puff Daddy. Jesus! It was almost like they couldn't see what was happening with hip-hop. Did they not really understand? For me, it was really important to be a part of that world. It was a challenge to do that track, and I enjoyed it."

Page and Diddy would perform the song together one more time, on October 9, 1999, at Giants Stadium in New Jersey for the charity NetAid, an antipoverty initiative. Page, with drummer Michael Lee (Page/Plant) and bassist Guy Pratt (Page/Coverdale), also treated the audience to two instrumentals, a new song entitled "Domino," and a wonderful arrangement of "Dazed and Confused" that featured a harmonically advanced reworking of the vocal melody played on the guitar.

WHEN IT WAS reported that Page was going to join forces and tour with the Atlanta-based Black Crowes in autumn 1999, it seemed like a bolt from the blue skies over Georgia. Truth is, there had been a rather lengthy courtship that preceded the announcement. In 1995, Robert Plant had brought Page to a Crowes show at London's Albert Hall, and a few nights later, in Paris, Jimmy got onstage to jam with the band. The two camps continued to cross paths over the next few years, giving their collaboration a sense of inevitability. It finally came together when Page was asked to spearhead a concert for two of his favorite charities: SCREAM (Supporting Children through Re-Education and Music) and ABC (Action for Brazil's Children).

"It was to take place at a London club called the Café de Paris," Page says. "Robert Plant and I had played the year before, and I wanted to do something different. My friend Ross Halfin suggested playing with the Black Crowes, who were in London at the time to play Wembley Stadium."

The Crowes were flattered by the invitation and immediately accepted. Zeppelin had been a huge influence on their music, and it wasn't difficult to see why Page was attracted to them. They both played a similar form of adventurous blues rock, and singer Chris Robinson sounded eerily like a blend of Steve Marriott and Terry Reid, two singers Page had once considered for Zeppelin.

On June 27, Page and the Black Crowes played a blistering ten-song show for the approximately four hundred people who packed the café. The set list was made up of blues standards including Jimmy Rogers's "Sloppy Drunk" and Elmore James's "Shake Your Moneymaker," along with a few Zep favorites like "Whole Lotta Love" and "Heartbreaker." The vibe was so good that, a couple of months later, Crowes manager Pete Angelus called Jimmy to see if he was interested in doing more shows. It sounded like fun to the guitarist. After exploring the more exotic and acoustic textures of his music with

Robert Plant, he was ready to play some unabashed hard rock and blues with the crackerjack Southern outfit.

Six shows were booked, including two that would be recorded at the Greek Theater in Los Angeles. In addition to playing Zeppelin concert favorites like "Celebration Day," "In My Time of Dying," and "Lemon Song," Page and the Crowes tackled rarely performed Zeppelin standouts like "Nobody's Fault but Mine," "Out on the Tiles," "Hots on for Nowhere," and the intricate "Ten Years Gone."

"When we did 'Ten Years Gone,' it was the first time I'd ever heard all the guitar parts from the record played live," Page says. "It was like being in guitar heaven. The great thing is they'd really done their homework. There was hardly any time where I had to say, 'Actually, it goes like *this.*' "

The band also performed a unique version of "Shapes of Things" that melded the Yardbirds version with the one recorded by the Jeff Beck Group. In homage to his old friend, Page played a note-perfect re-creation of Beck's wild Yardbirds solo, something he hadn't attempted since 1969.

But it wouldn't have been a true Page endeavor without his incorporating some element of innovation. It was decided that the double album *Live at the Greek: Excess All Areas* would be distributed exclusively on the Internet, through musicmakers.com. Considering that, at that time, iTunes was still just a glint in Steve Jobs's eye, this was an incredibly controversial and forward-looking move. Music-industry skepticism melted away, however, when Page and the Crowes became the first artists ever to crack the Top 10 with an Internet-only single, "What Is and What Should Never Be."

As Pete Angelus told journalist Alan di Perna, "It didn't take a genius to recognize that something very special was happening up there on the stage. Jimmy's manager, Bill Curbishley, and I both felt we'd be missing a really rare opportunity if we didn't record the shows to see what came of it. Initially, it was just, 'Let's get it on tape and see what happens.' Later on, Bill and I started talking about how this might be something that we wanted to release and how we could do it in a special way. I said, 'What avenue would provide us with the immediacy to get the music to the fans right away?' The Internet made sense on that level. Had we gone through a traditional major-label marketing-and-distribution system, I think it's safe to say that it would have been four to six months before the record would have been in stores."

Another factor that made an independent, Internet-only release possible was the fact that, at the time, neither party had a record deal. The Crowes had just left Sony, and Page was not signed as a solo artist, so both were free to do as they wished—with one exception. "In the Sony contract," Angelus explained, "there was a two-year holdback saying that the Black Crowes couldn't re-record any songs of theirs that had been commercially released. And that's the reason there were no Black Crowes songs on *Live at the Greek.*"

Upon its release, the album became the best-selling music product in the history of the Internet. *Live at the Greek* did not trigger the music-industry stampede toward the Internet, but it certainly made record labels stand up and take notice. It also reestablished Page's reputation as an artist at the vanguard of the music business.

Unfortunately, as was the case with his other post-Zeppelin ventures, his promising and powerful collaboration with the Crowes did not last long. A short tour to promote the album, with the Who as coheadliners, was cut short when Page became sidelined with back problems and was never to be resumed.

THE NEXT FEW years were relatively quiet but far from unproductive, as Jimmy labored for months on two very significant Led Zeppelin–related projects. The year 2003 saw the release of the *Led Zeppelin* DVD, which featured the only live performances of the band to have been professionally filmed during its twelve-year lifetime (described in detail in chapter seven).

Complementing the DVD was *How the West Was Won*, a significant three-CD set culled from Zeppelin's 1972 shows at the Los Angeles Forum and Long Beach Arena on June 25 and 27, respectively. These performances were legendary among Zep scholars and bootleggers, and you can hear why. Along with inspired renditions of "Immigrant Song" and "Stairway to Heaven" are a powerful, twenty-three-plus-minute version of "Whole Lotta

Love" and a positively orgasmic, twenty-five-minute performance of "Dazed and Confused."

As Page explains, "The L.A. shows just came zooming out of the speakers, and you could tell we were right on top of it. When we played like that, there was almost a mysterious fifth element at work."

Both compilations were enthusiastically received. *How the West Was Won* debuted at number one on June 14, 2003, becoming the first Led Zeppelin album since 1979's *In Through the Out Door* to do so. But the DVD went ballistic. The RIAA certified the video documentary at twelve times platinum, and according to the BBC it broke all sales records for a music video, selling nearly three times more copies in its first week than its nearest competitor. It remained the best-selling music DVD in America for three years.

In 2005, Page was appointed Officer of the Order of the British Empire (OBE) in recognition of his work in behalf of the charities Task Brazil and Action for Brazil's Children's Trust, and was made an honorary citizen of Rio de Janeiro later that year. His interest in the welfare of Brazil's poverty-stricken children was an outgrowth of a serious relationship he had with a native of that country, Jimena Gomez-Paratcha, whom he met there while on the *No Quarter* tour with Plant. He adopted Gomez-Paratcha's daughter, Jana, and the couple went on to have two children: Zofia Jade in 1997 and Ashen Josan in 1999.

In September 2007, Page's break from the spotlight ended with the earthshaking news that Led Zeppelin would reunite for one night, on November 26, at London's O2 Arena. Page, Plant, and bassist John Paul Jones would take the stage with Jason Bonham, son of John, to headline a concert in honor of Atlantic Records founder Ahmet Ertegun, who'd died the previous December. The show would represent the first time Led Zeppelin's founding members had performed together since May 1988, when they played at Atlantic Records' Fortieth Anniversary concert, also with Jason Bonham on drums.

To add to the excitement, the 1976 concert film *The Song Remains the Same* was reissued just days before the November 20 show as a deluxe two-DVD set. Featuring performances from the band's three-night stint at

Madison Square Garden in July 1973, the film was remixed and remastered in 5.1 Dolby Digital surround sound for the DVD. The package also included more than forty minutes of previously unreleased material, including performance footage of "Over the Hills and Far Away" and "Celebration Day."

While the reissue of *Song* may have struck some as an opportunistic way to cash in on the reunion show, the two were linked thematically: Both the O2 performance and the revamped concert film gave the band an opportunity to address some long-standing unfinished business. First there was the matter of Led Zeppelin living up to their legacy. During their eleven-year reign at the top of the rock world, Zeppelin went from strength to strength, producing one of the music's most durable bodies of work. But those interested in discovering chinks in the band's armor usually cited the band's two previous reunions: the 1985 appearance at Live Aid, featuring drummers Tony Thompson and Phil Collins, and the 1988 set at the Atlantic Records Fortieth Anniversary concert.

Even Page admits that both shows, for all the excitement they generated, had been disappointing. "The reunion at the O2 Arena represented an opportunity to finally present Zeppelin properly, and we took it very seriously," he says. "The performances at Live Aid and the Atlantic event were not good for various reasons. That wasn't going to be the case at the O2 show."

Second, there was the matter of *The Song Remains the Same*, which had been in great need of an overhaul for some time. As Page notes, when the film and soundtrack entered the digital realm in the nineties, they "never received the care they deserved." The new DVD and CD reissues did more than simply do justice to the originals. Page actually improved upon their sound by employing the cutting-edge tools of modern audio technology.

The O2 show was everything the band and twenty thousand fans had hoped for. The reconstituted and well-rehearsed Zeppelin played sixteen songs that ran the stylistic and chronological gamut of their entire career, including two numbers they'd never played in their entirety in concert: "Ramble On" and "For Your Life."

"Initially, they asked us to play a certain amount of time, but we extended it to get more songs in," Page says. "We quickly realized that we couldn't play 'Whole Lotta Love' for thirty minutes, have a drum solo, and then play 'Stairway to Heaven' for twenty minutes and leave! You know, do 'Rock and Roll' as an encore and be off—we just couldn't do that. So in order to show people how we used to perform, and play with flair and passion, we had to play a pretty long set."

Highlights of the show included an electrifying "Dazed and Confused," with Page wielding his bow with an even darker magic than he had in the seventies. The rendition of "Kashmir," with its windswept grandeur, was nothing less than epic. But Page, who let his hair go a wizardly white for the event, reserved what was perhaps the most soul-stirring—and surprising—moment of the show for the end of "Stairway to Heaven." Despite its widely accepted status as the "greatest rock guitar solo of all time," Page rarely played the "Stairway" break note for note in concert, preferring to improvise around the themes heard on the *Led Zeppelin IV* version. But he did so at the O2 show, bringing down the house in the process. "I don't think anybody thought I could actually play it!" he says with a laugh. "I guess I just wanted to show I could."

Perhaps the only disappointing aspect of the O2 reunion was that it was limited to just one show. With more than just a hint of exasperation still evident years after the fact, Page explains: "Robert didn't want to continue. It's a bit silly, because there was such a massive demand. It was selfish to do just one show. To some degree, I'm not sure if we should've taken the genie out of the bottle if we weren't going to follow through."

WITH THE GENIE firmly sealed, Page forged ahead with a quartet of high-profile projects aimed at four different media: television, film, print, and the Internet. It was no small achievement for an artist approaching his late sixties. First up was his performance of "Whole Lotta Love" at the 2008 Beijing Olympics' closing ceremony, which was viewed on television by an estimated two billion people across the globe. Next came *It Might Get Loud*, Davis Guggenheim's 2009 documentary that explored the electric guitar through the eyes of Page, the White Stripes' Jack White, and U2's the Edge. In September 2010, Page released *Jimmy Page by Jimmy Page*, a limited-edition "photo autobiography" conceived and codesigned by the guitarist and published by Genesis Publications. Finally, on July 14, 2011, Page launched his own website, jimmypage.com, intending it as a vehicle to present his "past, present, and future work."

According to Page, the closing ceremonies of the Beijing Olympic games were nothing less than "awe-inspiring." For the performance of "Whole Lotta Love," he was paired with pop singer Leona Lewis, winner of the UK's popular talent show program *The X Factor*. Accompanied by bursts of fireworks and hundreds of dancers, acrobats, and drummers, Page and the sultry vocalist arrived at Beijing's National Stadium, nicknamed the Bird's Nest, atop a red double-decker bus, from which they delivered a rocking if bowdlerized version of "Whole Lotta Love," whose sexually explicit lyrics and moans were altered in order to meet the requirements of official Chinese censors.

Page's segment was part of an eight-minute entertainment extravaganza marking the handover of the games to London, host of the 2012 Olympics. "I just really loved doing something like that, because it shows you how music can reach so many people," Page says. "That performance was seen all over the world and in every province of China. It was a great display of color, spectacle, and drama—Leona sang it in such a ballsy way—but the true beauty of it was that a great rock and roll riff powered the whole thing.

"Who knows how that resonated in China?" Page says. "Rock and roll helped change Russian society in many ways when the wall came down. A lot of people in China came up to me afterwards and told me how much they enjoyed the performance, and I thought, Yeah, mission accomplished! Fucking great! There it is, after everything—basically it comes down to that riff, you know?"

Traveling with Page was his friend and photographer Ross Halfin, who recalled that memorable performance.

"The ceremonies included the handing over of the Olympic flag to London's mayor, as the city was next in line to host the Olympics," Halfin says. "Jimmy was one of a few select people chosen to represent the UK at the handover. While he wasn't going to be performing with Zeppelin, it was a huge event and an opportunity for me to photograph Jimmy on a historic occasion.

"We flew out a week earlier to rehearse and get acclimated. The flight arrived at five thirty A.M. at a super-modern terminal that made all our Western airports look antiquated. We were met by a student holding up a sign and told to follow him to immigration and customs. It was a bit like a European school trip.

"Outside, the heat blasted us; it was already ninety degrees. We drove an hour into central Beijing, the odd thing being that there was no traffic on the road. We arrived at our hotel, the Grand, which was surrounded by barricades, barbed wire, and what looked like half of the Chinese national police force. We all had to line up again and go through lots of airport-type security far more stringent than you have going through JFK or LAX. We eventually got to our hotel, and the odd thing was that there appeared to be no one there except for us and a few Olympic officials. It was literally like something out of *The Twilight Zone*. Unlike the ultramodern airport terminal, the hotel looked like it hadn't been updated since 1970.

"I decided to go for a walk back through the two lots of security, into huge empty streets. I walked about a mile and came to more barricades, with armored cars, tank traps, and so on. On the other side was normal China: people shopping, eating, drinking, and doing everyday things. It was like we were trapped in an Olympic bubble.

"Rehearsals for the handover show were held in an old airfield past the Great Wall, an hour away. For each of the next four days, Jimmy and I drove out and spent the whole day there, all so he could do about five minutes' worth of work each time. We went out a couple of times in the evening, but it was so tedious going through security that we mostly stayed in the empty hotel. On one of my walks out of our luxury prison, I found the Grand Hyatt. It was a normal hotel with Chinese people and none of the crap attached to it. I told Jimmy and we immediately transferred, much to the chagrin of the Olympic committee.

"On the day of the event, Jimmy and I, along with Leona and footballer

David Beckham—who brought along an entourage of apparently several thousand bodyguards—squeezed onto a bus and headed to Beijing's Bird's Nest stadium for the ceremony. It was 110 degrees outside and there was no air-conditioning in the bus. We stopped three times on the way to the stadium, and then went through the usual two rounds of security and walked half a mile to a compound where there was one final rehearsal. We were inside the Olympic complex and at least a thirty-minute walk from the stadium. No toilets, no catering, no nothing. The Olympic people really do know how to look after you!

"After hanging around for another half a day, we were eventually led into the bowels of the stadium. Finally, at eight thirty P.M., Jimmy, Leona, and David got on the 'Magic Bus'—a traditional London red double-decker that drove them out onto the stadium field. The audience erupted as the bus arrived, and Jimmy launched into the opening riff of 'Whole Lotta Love.' Leona did an excellent job, and I found photographing the show from the field quite nice and easier than shooting a gig.

"Due to the heavy demand for flights out of China, Jimmy and I decided to stay a couple more days. It turned out to be the best part of the trip. Everywhere we went, everyone, old and young, recognized Jimmy—with all of them playing imaginary guitars! One morning, at seven A.M., we went to the Forbidden City. It was packed, and Jimmy was mobbed.

"Even leaving China was a pleasurable experience. There was no hassle at the airport; all the immigration and customs people wanted was a picture and autograph from their new guitar god. Jimmy was most willing to indulge them."

WHILE PERHAPS NOT as monumental as the Beijing Olympics, Page's next endeavor was equally fascinating and very impressive. Producer Thomas Tull, president and CEO of Legendary Pictures, and director Davis Guggenheim decided to realize a dream they had to "capture the beauty of the guitar on film." Given that Tull had produced such blockbusters as *Batman*

Begins, 300, and *The Hangover,* and that Guggenheim had won an Academy Award for his groundbreaking, Al Gore–narrated documentary on global warming, *An Inconvenient Truth,* their dream had teeth.

Instead of producing a straightforward history of the guitar, it was Tull's idea to select three rock guitar players from different generations to explain, in their own words, the importance of the instrument to their art and its impact on the general culture. Topping Tull and Guggenheim's wish list were Page, the Edge, and White.

"We probably wouldn't have made the movie if we couldn't get those guys," Guggenheim said. "This was the movie we wanted to make, and we were determined to do it right."

The concept of the film was simple and appealing. The first part of it was built around three self-contained segments in which each individual artist spoke of his personal relationship with music and the guitar. This was followed by an unrehearsed "summit" meeting, where the three played each other's songs together.

"Davis, who had just done the Al Gore film, contacted me and outlined the project," Page says. "He was obviously a music fan and I liked that. He had passion. And one thing he said was, 'First, we'll have an interview and I'll record it. It won't be on-camera, though, but more of a get-to-know-you thing, and to build some momentum.' And I thought, Hey, that's cool. The whole thing grew out of that."

One of the more interesting moments in the film features Jimmy showing the other two guitarists how to play the central riff of "Whole Lotta Love." For a moment, White and the Edge visibly morph into their thirteen-year-old guitar-obsessed selves, giddy at the fact that they're about to learn one of rock's iconic riffs from its equally iconic creator.

As White explained, "That riff is one of those things you grow up with. It's embedded in you like a nursery rhyme." The Edge concurred: "A song like 'Whole Lotta Love,' we know it so well—it's like the Bible or a street sign. But to see the original fingers playing it . . . it's like going inside the pyramids."

It Might Get Loud was warmly received by critics and was a box-office success, eventually becoming one of the one hundred highest-grossing documentaries of all time. While it was not quite the epic statement on the guitar Tull and Guggenheim set out to make, it was a captivating record of three historically important musicians. For many, it would be the first time they would have an opportunity to experience Page, the Edge, and White up close and personal.

T HE FIGURATIVE PAINTER Francis Bacon once remarked, "The job of the artist is to always deepen the mystery." Nobody in the music world had understood this maxim better than Page, who for years was the paradigm for rock-star inscrutability. But his most recent activities demonstrate a new desire by him to have his achievements appreciated and understood. If *It Might Get Loud* was his first attempt at explaining his art, his next project went one step further. For the previous few years, Page had been painstakingly poring over thousands of photos, verifying dates and timelines in order to create what he called a "photographic autobiography." Entitled *Jimmy Page by Jimmy Page* and published by Genesis Publications, the massive, leather-bound volume represented the most complete portrait of the rocker to date, featuring more than 650 images—some from the guitarist's personal collection—carefully annotated by Page.

"The book tells the story of my life as a musician," Page explains. "It was designed to show where my passion for music started and how it evolved. But at the same time, I wanted it to be evocative. I looked for pictures that had subtle connections and little points of reference that you won't notice straightaway but will pick up on after repeated viewing. The truth is, nobody else could have made this book."

Genesis publisher and designer Catherine Roylance agrees. "He was involved in every detail and was really passionate about the content, the binding, and all the materials," she says. "He has a great eye, and he could see the play of images together and the pace of the book from one page to the next. I don't think this sort of 'photographic autobiography' has ever been done before, which makes it a landmark publication."

The book unfolds in strict chronological order, beginning with an astonishing series of shots from the late fifties that capture a young Page rocking out with his early guitars (a Hofner President and Grazioso Futurama) and his early bands (Red E. Lewis and the Red Caps and Neil Christian and the Crusaders). Rare glimpses of his life as a studio musician in the mid-sixties are followed by a number of images of Page with the Yardbirds, including a fabulously intimate picture of him and Jeff Beck tuning up before a modest gig at the Staples High School auditorium in Westport, Connecticut.

Of course, the Led Zeppelin years are amply represented, featuring never-before-seen shots of the band at every stage of its career. The book also makes fascinating stops at locations like Bron-Yr-Aur in Wales, where Page and Plant composed songs for the band's third album. Another photo, taken in 1971, depicts Page standing in front of his rarely seen Boleskine House in Loch Ness, once owned by Aleister Crowley.

The book also does justice to Page's post-Zeppelin years, including photos of the guitarist performing on the ARMS tour with Eric Clapton and Beck, playing the Cambridge Folk Festival with Roy Harper, jamming in Marrakech, and closing out the 2008 Summer Olympics in Beijing. It all adds up to an illuminating look at the career of the enigmatic guitarist.

While *Jimmy Page by Jimmy Page* is revealing, it also reflects Page's natural reluctance to share *too* much of himself with the world. He volunteers very little about his personal life, his commentary rarely goes beyond two or three sentences, and the book itself was limited to a run of 2,500 copies—each bearing a price tag of approximately $700, guaranteeing that its readership would be limited to a small group of hard-core, well-heeled fans.

Perhaps as a gesture to his millions of other fans, his next project would be universally accessible. Inspired by his research for the Genesis book, Page felt it was time for him to offer something similar on the Internet, and JimmyPage.com was born.

"I'd had the domain name for a number of years, and I was just sitting on it," he says. "I felt it was the right time to put something together."

One factor that led to his taking the Internet plunge was the existence of multiple Jimmy Page websites: "When you've got fans, and especially ones

who are committed to doing websites, it's okay," Page says. "The only problem I have is when inaccuracies pop up and gain a life of their own. Over the
years, with the Internet and forums, urban musical myths about me and my
career are now seen as the truth.

"Another reason for doing the website is, if you talked to people who
aren't switched on to what I've done, they probably think, 'Oh, yeah, he was
the guitarist in Led Zeppelin, wasn't he? And they did the o2.' And if they
said anything else about me, it would probably relate to that BBC clip [of
Page performing on Huw Wheldon's *All Your Own* in 1957], where I'm about
thirteen. That's all they really know about me—and I've been seriously active for over fifty years."

To date, the most interesting and ambitious section of the site has been
the opening splash page, which changes every day to coincide with some
event in Page's life. Much like *Jimmy Page by Jimmy Page*, the splash page
features an interesting photo with a three- or four-sentence introduction
written by the guitarist explaining its importance. The twist is that the images are often accompanied by rare bits of audio or video clips from Page's
personal archives that complement or amplify the significance of the date.
While much of the material is quite good—including a surprisingly strong
performance from his Neil Christian days and an amazing "In My Time of
Dying" from his *Outrider* tour—much of the information is tantalizingly
scant, leaving visitors to the site hungry for more. "Deepen the mystery,"
indeed.

MUSICAL INTERLUDE

—————

A CONVERSATION WITH MEN'S FASHION
DESIGNER JOHN VARVATOS

———————

MOST ENTERTAINERS ARE lucky to create a single iconic image during their careers. Jimmy Page, however, created many: the pre-Raphaelite dandy with a kaleidoscopic Telecaster; the black-and-silver-star-spangled arena showman wielding a sunburst Les Paul; and the decadent, dragon-suited Dark Lord of the Gibson double-neck are just a few of his immediately recognizable personas, and they continue to resonate with today's leading fashion experts.

John Varvatos, one of the world's leading contemporary clothes designers, cites Page as one of the single most influential figures in rock and roll chic. "Very few musicians have had an impact on fashion that transcended their generation," says Varvatos, who has designed for Calvin Klein and Ralph Lauren and now runs his own successful fashion house. "Jimmy is on that very short list. The looks that he pioneered in the sixties, the seventies, and beyond still have an effect on designers all over the world. We are constantly being stimulated by those classic images of him with Led Zeppelin, and he has influenced the way we think. The look he cultivated throughout his career was really thought out, and it was perfection. It was really perfection."

What makes Jimmy Page's style unique?

JOHN VARVATOS Some musicians have a personal sense of style, but there aren't many. Most never really understand. Jimmy was about the whole look, down to what kind of scarf would look good. He definitely understood stage clothes, too. He knew how to look good close up and from far away, and he knew how to make himself look big onstage. Even when he grew a beard, it didn't look shabby or scraggly—it looked cool. It just worked. My other big fashion influence was the actor Steve McQueen, who had a similar thing. There's an aura about them that lets them carry it off.

Jimmy still looks great. Letting his hair go white was a perfect example of doing the right thing. He looks in command. There's an aura about him—there's always been an . . . *aura*.

How did Page influence your work?

I've always had a passion for music and for fashion, but Led Zeppelin is what connected them. Even now, if you look at my inspiration boards for both men and women, they often have images of Led Zeppelin and Jimmy Page attached to them. It's not about copying a look; it's about capturing a vibe. I particularly liked that early-seventies period where he got dressed up, but in a really funky way.

When people think of style in the seventies, they usually mention David Bowie or Marc Bolan. Jimmy isn't an obvious choice.

In the seventies, style became very androgynous. People like Lou Reed, Bowie, and Bolan were dressing in a more feminine way, but Jimmy never crossed the line. He dressed in the style of the day but never got wrapped up in it. He had his own point of view; he walked to his own beat, and he wasn't a follower.

Musicians are in touch with both male and female energy. Somehow Jimmy cut it just right, so that his clothes were flamboyant but still very masculine. I was looking at some photos recently of the band UFO that were taken in the seventies. It was clear that they were influenced by the

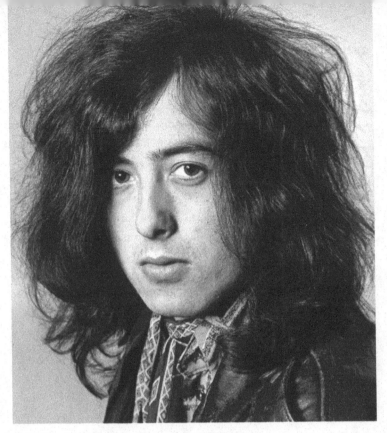

Early Led Zeppelin promo shot, 1968 (© *Dick Barnatt/Getty Images*)

androgyny of the times, and some of the members were actually wearing women's tops. I never felt that Jimmy was wearing women's clothes.

Perhaps it was because his color palette was relatively understated.
He understood the edge to dressing in black—the power of it. It's something a lot of people discovered over time, but he knew that from the beginning. Jimmy had a narrow color palette, but he stretched it in a really tasteful way by embellishing his basic black-and-white outfits with the dragon and embroidered symbols.

The really great thing was that he never looked like he was doing theater; he looked like he was going to play a rock show. Many musicians can't take what they have in their fingers and paint that other picture. With Jimmy, you know he was always looking at the bigger picture. What will the stage look like? What will these people look like on that stage? He was clearly someone who is obsessed with detail, right down to his using a violin bow on his guitar.

Is Jimmy Page's style still relevant?

It has really had an effect on fashion for over thirty-five years. Even if it's not a literal thing, the fashion world is still constantly stimulated by those images.

Jimi Hendrix had his impact—his military jacket is still prominent in the fashion industry. Young people still look at Hendrix and think he's cool, but Jimmy Page is a much bigger thing.

GRAND FINALE

THE ASTROLOGY OF JIMMY PAGE AS INTERPRETED BY NOTED
STARGAZER MARGARET SANTANGELO.

JIMMY PAGE'S INTEREST in astrology is no secret. You could say he wore it on his sleeve—or, to be more precise, his pants. While his exact birth time on January 9, 1944, is up for debate, the most important elements of his birth chart were embroidered on the left hip of the dragon suit, the black outfit he wore onstage in 1975 and 1977. The glyphs depicted Capricorn for his sun sign, Scorpio for his rising sign, and Cancer for his moon sign. These three signs are the most dominant and interesting astrological archetypes in Jimmy Page's horoscope.

By emblazoning his performance attire with emblems and talismans of esoteric significance such as his astrological makeup, Page was doing more than making an eccentric fashion statement: He was evoking and making manifest the subtle qualities that these symbols represent, and both consciously and subconsciously reinforcing these key aspects of his individual essence.

The sun sign is the representative of the ego in the birth, or natal, chart, which shows the position of the stars at an individual's time and place of birth. Sun in Capricorn bestows ambition, practicality, and business savvy. In Jimmy Page's chart, it falls in the second house of money, values, and finances. This sun position ensures that Jimmy would never have any trouble earning money. In fact, this indicates that he would be quite an astute businessman.

"I financed and completely recorded the first album before going to Atlantic [Records]," Page said. His financial acumen, as well as his ultimate profitability and commercial success, is a second-house Sun-Mercury conjunction at work. His ability to use the marketplace as an outlet for his creative output while maintaining a sense of artistic integrity can be credited to the disciplined, shrewd, and cautious Capricorn influence.

The second glyph on Page's dragon suit represents Scorpio on the ascendant, and it may be the most important. The ascendant, or rising, sign indicates how you project yourself to others—that is, how you are seen, as opposed to who you actually are deep down inside (which is a function of the sun). Jimmy Page's persona—as well as what he projected to the media and the outside world—was from the very beginning that of a Scorpio. In fact, most aspects of his persona for which he is known—his interest in the occult, his secrecy and privacy, his controlling nature—are qualities of his rising sign in Scorpio. The Scorpio ascendant gave Jimmy a place to hide his true self, represented by the Capricorn sun, and allowed him to project a more subtle, obscured, and nuanced persona to the world.

The last glyph represents Cancer, the fourth sign of the zodiac. Jimmy's own moon falls in Cancer, and since the moon is the natural ruler of Cancer, this is an especially powerful position. It indicates a very intensely emotional person, one who is often very intuitive and perhaps even psychic. Moon in Cancer can also defer depressive or escapist tendencies and is often found in the charts of musicians who overindulge in alcohol or drugs. The moon in Cancer is the vehicle through which Jimmy accessed his deepest emotions and thoughts and channeled them into his music. It is also the source of his self-destructive tendencies, namely his tendency toward substance abuse.

Intensifying this lunar influence, his Cancer moon falls in the Scorpio eighth house, the domain of shared resources, sex, and all that is hidden. This is evidenced by his self-protective and reclusive qualities. With so much Scorpio energy combined with a large number of retrograde planets, it is a wonder that he ever made it out of the studio and in front of an audience.

Page's dragon pants with astrological symbols (© *Kazuyo Horie*)

However, Jupiter in showy Leo in the ninth house, just shy of his midheaven, ensured that he would not just perform but also give audiences one of the greatest shows on earth. Moreover, Jupiter—the good-luck planet—is the highest planet in his chart, indicating that he would be very successful professionally.

Another persistent theme in Jimmy's chart is the retrograde. At the time of his birth, every planet except Venus was retrograde. This influence creates a person who is very internally focused, uncomfortable in the public eye, and naturally introspective. With Mercury retrograde exactly conjunct with his sun in Capricorn, spirituality and self-reflection come naturally to him. This confers great importance on Venus in Sagittarius, the sign of spirituality, intellect, higher education, publishing, and foreign travel, which falls in his first house. This indicates that, despite the introverted energy of all the remaining planets, Venus is a primary indicator of how exactly he would make his mark on the world and assert his ego, which is the primary function of the first house, in which Venus falls.

One such Sagittarian/ninth-house venture was Page's occult bookstore, Equinox, which he opened in London in 1973. Page did not expect to make money from the bookshop but "basically wanted the shop to be a nucleus, that's all," as he said. He did publish a book translated by Crowley, *The Book of Goetia*, as well as provide a center for the occult social network in London.

In addition, Venus in Sagittarius prompted Page to become a distributor of both his music and that of fellow musicians by creating the label Swan Song, also in 1974. The strength of Page's Venus, combined with his Capricorn mastery of business, enabled him to launch this highly successful label, which released records by Led Zeppelin and by artists handpicked by Page, such as Bad Company and the Pretty Things.

Venus is one of the most important planets in any artist's horoscope, representing the voice, singing, and music in general (as well as all the creative arts). In Page's chart, it is the only direct planet and it is in the first house of ego and identity, in Sagittarius, which rules spirituality, foreigners, travel, education, and philosophy. Page's reverence for music and his almost religious zeal and dedication to all aspects of the creative process illustrate the energy of Venus in Sagittarius. Because Page considered creating music to be a transcendental endeavor, he was able to subjugate the ego demands of the first house in order to foster a more harmonious band dynamic. Page used metaphysical terms to describe his natural, intuitive synthesis with bandmate Robert Plant, noting that "the two of us were just channeling the music. . . . It was almost effortless."

Page, through the work of Led Zeppelin, explored the metaphysical aspects of music, both communing with and communicating through the spirit in a mystical quest to find the light of wisdom by means of music. Page is, in fact, a musical alchemist: Through the manipulation of sound, frequencies, technology, and whatever else was needed, he actualized a resonant, living manifestation of the very essence of the ethereal creative spirit.

However, the Hermit is not only the seeker of the light—he is also a bearer of the light. Page, throughout this journey, has overcome the potential limitations of his reticent, mysterious Scorpionic tendencies and emerged from the shadows to share the secrets of his sonic sorcery. Neptune, the planet of intuition, spirituality, and cosmic creativity, is currently opposing Page's midheaven, prompting him to complete his transformation into that timeless archetype which so resonated with his young mind: that of the wise, learned sage shining his lantern of wisdom from on high to the unenlightened initiates who would follow in his footsteps.

THANK YOU

First and foremost, I'd like to thank Jimmy Page for his time and patience over the years. Whenever I interviewed him it always went something like this: The first hour was always great, the second hour is when we got into a groove, and the third hour was when he politely started looking for the nearest exit. But God bless him, he always slipped away politely.

Next, I would like to thank my favorite rock and roll photographer, Ross Halfin, whose work graces this book, and his lovely assistant, Kazuyo Horie. Ross takes the best pictures of Jimmy, and he and Kaz are wonderful company whenever I visit London.

On several occasions, the energetic Mr. Greg Di Benedetto served as my wingman on interviews with Jimmy. A great guitarist in his own right, Greg filled in gaps while I would sometimes desperately search for the next relevant thread of questioning, and for this, I thank him.

Next I would like to bow to Alexis Cook, Chris Scapelliti, Alan di Perna, Harold Steinblatt, and John Bednar. Alexis is the brilliant art director who helped me design the interior of this rather complicated book, Chris functioned as my closest advisor, and Alan willingly allowed me to plunder and pillage some of his best work on Page's solo years. Mr. Steinblatt also made a number of invaluable eleventh-hour edits and suggestions, while guitar tech supreme Bednar schooled me on Jimmy's gear. There is a special place in heaven reserved for all of you.

For supporting my work despite the fact that I hogged the kitchen for a good year with my computer and research, I offer kudos to my wife, Lorinda, my son, Kane, and my daughter, Nico.

Finally, thanks to Phyllis of the Thelesis Lodge of Philadelphia chapter of the Ordo Templi Orientis, Chris Dreja, John Varvatos, Paul Rodgers, my superfab agent David Dunton, my superfab editor Charles Conrad, Joe

Bosso, Margaret Santangelo, Peter Makowski, Dave Brolan, Bill McCue, Steve Karas, Jaan Uhelszki, Izzy Zay, and the *Guitar World* crew of Anthony Danzi, Jimmy Brown, Tom Beaujour, and Jeff Kitts, who allowed me to play hooky during the deadline for this book. I couldn't have done it without you!

BIBLIOGRAPHY

BOOKS

Case, George. *Jimmy Page: Magus, Musician, Man*. Hal Leonard, 2007.

Clayson, Alan. *The Yardbirds*. Backbeat Books, 2002.

Crowley, Aleister, ed. *The Goetia: The Lesser Key of Solomon the King*, trans. Samuel Liddell MacGregor Mathers. Weiser Books, 1995.

Daniels, Neil. *Robert Plant: Led Zeppelin, Jimmy Page and the Solo Years*. Independent Music Press, 2008.

MAGAZINE ARTICLES

Behutet Editors. "Interview: Eric Hill, Manager of London's Equinox Book Shop." *Behutet* 37, Vernal Equinox 2008.

Blake, Mark. "Graffiti Art." *Guitar World*, May 2005.

Bosso, Joe, with Greg Di Benedetto. "Physical Riffiti." *Guitar World*, January 1991.

Burroughs, William S. "Rock Magic." *Crawdaddy*, June 1975.

Di Perna, Alan. "Birds of a Feather." *Guitar World*, May 1995.

———"Higher Ground." *Guitar World*, June 1998.

Harper, Clive. "The Equinox Bookshop: Some Images and Impressions." *Behutet* 33, Vernal Equinox 2007.

Houghton, Mick. "The Jimmy Page Page." *Sounds*, July 10, 1976.

Icenine, Alexander. "Led Zeppelin III." *Creem*, December 1970.

Ingham, John. "Technological Gypsy." *Sounds*, March 13, 1975.

Makowski, Peter. "Speak of the Devil." *Guitar World*, Holiday 2006.

Mendelsohn, John. "Led Zeppelin I." *Rolling Stone*, March 1969.

———"Led Zeppelin II." *Rolling Stone*, December 1969.

Rosen, Steven. "Jimmy Page: The Interview." *Guitar World*, July 1986.

WEBSITES
www.jimmypage.com
www.ledzeppelin.com
www.led-zeppelin.org
www.tightbutloose.co.uk
www.wholelottaled.webs.com

Brad Tolinski has been the editor in chief of *Guitar World*, the world's best-selling magazine for musicians, for over two decades. He has interviewed and profiled most of popular music's greatest guitarists including Eric Clapton, B. B. King, Edward Van Halen, Jack White, and Jeff Beck. In addition to *Light & Shade: Conversations with Jimmy Page*, he has written two deluxe-edition art books for Genesis Publications in England, *Classic Hendrix: The Ultimate Hendrix Experience* and *The Faces: 1969–75.*

This "oral autobiography" of Jimmy Page, the intensely private mastermind behind Led Zeppelin, is the most complete and revelatory portrait of the legendary guitarist ever published.

More than thirty years after disbanding in 1980, Led Zeppelin continues to be celebrated for its artistic achievements, broad musical influence, and commercial success. The band's notorious exploits have been chronicled in bestselling books, yet none of the individual members of the band has penned a memoir nor cooperated to any degree with the press or a biographer. In *Light & Shade*, Jimmy Page, the band's most reticent and inscrutable member, opens up to journalist Brad Tolinski, for the first time exploring his remarkable life and musical journey in great depth and intimate detail.

Based on extensive interviews, *Light & Shade* encompasses Page's entire career, beginning with his early years as England's top session guitarist when he worked with artists ranging from Tom Jones, Shirley Bassey, and Burt Bacharach to the Kinks, the Who, and Eric Clapton. Page speaks frankly about his decadent yet immensely creative years in Led Zeppelin, his synergistic relationships with band members Robert Plant, John Bonham, and John Paul Jones, and his notable post-Zeppelin pursuits. While examining every major track recorded by Zeppelin, including "Stairway to Heaven," "Whole Lotta Love," and "Kashmir," Page reflects on the band's sensational tours, the filming of the concert movie *The Song Remains the Same*, his fascination with the occult, meeting Elvis Presley, and the making of the masterpiece *Led Zeppelin IV*, about which he offers a complete behind-the-scenes account. Additionally, the book is peppered with "musical interludes" that feature conversations between Page and other guitar greats, including his childhood friend Jeff Beck and hipster icon Jack White.

Through Page's own words, *Light & Shade* presents an unprecedented first-person view of one of the most important musicians of our era.